Springer Theses

Recognizing Outstanding Ph.D. Research

For further volumes:
http://www.springer.com/series/8790

Aims and Scope

The series "Springer Theses" brings together a selection of the very best Ph.D. theses from around the world and across the physical sciences. Nominated and endorsed by two recognized specialists, each published volume has been selected for its scientific excellence and the high impact of its contents for the pertinent field of research. For greater accessibility to non-specialists, the published versions include an extended introduction, as well as a foreword by the student's supervisor explaining the special relevance of the work for the field. As a whole, the series will provide a valuable resource both for newcomers to the research fields described, and for other scientists seeking detailed background information on special questions. Finally, it provides an accredited documentation of the valuable contributions made by today's younger generation of scientists.

Theses are accepted into the series by invited nomination only and must fulfill all of the following criteria

- They must be written in good English.
- The topic should fall within the confines of Chemistry, Physics and related interdisciplinary fields such as Materials, Nanoscience, Chemical Engineering, Complex Systems and Biophysics.
- The work reported in the thesis must represent a significant scientific advance.
- If the thesis includes previously published material, permission to reproduce this must be gained from the respective copyright holder.
- They must have been examined and passed during the 12 months prior to nomination.
- Each thesis should include a foreword by the supervisor outlining the significance of its content.
- The theses should have a clearly defined structure including an introduction accessible to scientists not expert in that particular field.

Shinsuke Inuki

Total Synthesis of Bioactive Natural Products by Palladium-Catalyzed Domino Cyclization of Allenes and Related Compounds

Doctoral Thesis accepted by
Kyoto University, Japan

Author
Dr. Shinsuke Inuki
Graduate School of Pharmaceutical Sciences
Kyoto University
46-29 Yoshida-shimo-adachi-cho
Sakyo-ku, Kyoto 606-8501
Japan
e-mail: shinsuke_inuki@fujifilm.co.jp

Supervisors
Prof. Hiroaki Ohno
Graduate School of Pharmaceutical Sciences
Kyoto University
46-29 Yoshida-shimo-adachi-cho
Sakyo-ku, Kyoto 606-8501
Japan

Prof. Yoshiji Takemoto
Graduate School of Pharmaceutical Sciences
Kyoto University
46-29 Yoshida-shimo-adachi-cho
Sakyo-ku, Kyoto 606-8501
Japan

Prof. Nobutaka Fujii
Graduate School of Pharmaceutical Sciences
Kyoto University
46-29 Yoshida-shimo-adachi-cho
Sakyo-ku, Kyoto 606-8501
Japan

ISSN 2190-5053 e-ISSN 2190-5061
ISBN 978-4-431-54042-7 e-ISBN 978-4-431-54043-4
DOI 10.1007/978-4-431-54043-4
Springer Tokyo Dordrecht Heidelberg London New York

Library of Congress Control Number: 2011940784

Cover design: eStudio Calamar, Berlin/Figueres

Printed on acid-free paper

Springer is part of Springer Science+Business Media (www.springer.com)

Parts of this thesis have been published in the following journal articles:

Shinsuke Inuki, Shinya Oishi, Nobutaka Fujii and Hiroaki Ohno, "Total synthesis of (±)-lysergic acid, lysergol, and isolysergol by palladium-catalyzed domino cyclization of amino allenes bearing a bromoindolyl group", Organic Letters, 10 (22), 5239–5242 (2008)

Shinsuke Inuki, Yuji Yoshimitsu, Shinya Oishi, Nobutaka Fujii, Hiroaki Ohno, "Ring-construction/stereoselective functionalization cascade: total synthesis of pachastrissamine (jaspine B) through palladium-catalyzed bis-cyclization of bromoallenes", Organic Letters, 11 (19), 4478-4481 (2009)

Shinsuke Inuki, Yuji Yoshimitsu, Shinya Oishi, Nobutaka Fujii, Hiroaki Ohno, "Ring-construction/stereoselective functionalization cascade: total synthesis of pachastrissamine (jaspine B) through palladium-catalyzed bis-cyclization of propargyl chlorides and carbonates", The Journal of Organic Chemistry, 75 (11), 3831-3842 (2010)

Shinsuke Inuki, Shinya Oishi, Nobutaka Fujii and Hiroaki Ohno, "Total synthesis of (±)-lysergic acid, lysergol, and isolysergol by palladium-catalyzed domino cyclization of amino allenes bearing a bromoindolyl group", Organic Letters, 13 (8), 2145 (2011)

Shinsuke Inuki, Akira Iwata, Shinya Oishi, Nobutaka Fujii, and Hiroaki Ohno, "Enantioselective total synthesis of (+)-lysergic acid, (+)-lysergol, and (+)-isolysergol by palladium-catalyzed domino cyclization of allenes bearing amino and bromoindolyl groups", The Journal of Organic Chemistry, 76 (7), 2072–2083 (2011)

Supervisor's Foreword

It is a pleasure to introduce Dr. Shinsuke Inuki's thesis work on application to the Springer Thesis Prize, as an outstanding original work in the world's top university. Dr. Inuki joined Prof. Fujii's group, Kyoto University, as an undergraduate student from April of 2005. In April 2006, he entered the Graduate School of Pharmaceutical Sciences, Kyoto University, and started his doctoral study with me at the same laboratory.

In recent years, catalytic cascade reactions have been recognized as an efficient approach to target molecules, by minimizing the number of steps and separation processes as well as the amount of time, labor, and waste involved. Dr. Inuki successfully applied the palladium-catalyzed domino cyclization of bromoallenes as allyl dication equivalents to cascade cyclization–stereoselective functionalization for asymmetric total synthesis of pachastrissamine (jaspine B), an anhydrophytosphingosine exhibiting antitumor activities. A short-step total synthesis of this natural product has been achieved by use of domino cyclization of propargyl carbonates. The thesis also describes his elegant synthetic work on total synthesis of ergot alkaloids based on palladium-catalyzed domino cyclization of amino allenes bearing a bromoindolyl group. The tetracyclic indole, the common synthetic intermediate for his ergot alkaloid synthesis, was directly constructed from the allenic substrates in a stereoselective manner. Using the key intermediate obtained, he achieved asymmetric total synthesis of lysergic acid, lysergol, and isolysergol.

It is noteworthy that all of these works were based on his very original ideas. The four outstanding papers, prepared by himself as the first author, have been published in the top journals in organic synthesis (*Organic Letters and the Journal of Organic Chemistry*). His total synthesis of lysergic acid was highlightened in *Synfact* (2009, 476) and *Organic Chemistry Portal* (2009, May 11).

His thesis study has shown that palladium-catalyzed domino cyclizations are useful for stereoselective construction of the core structures of natural products. These results would contribute to the synthetic and SAR studies of sphingolipids and indole alkaloids. I hope his outstanding thesis will contribute to synthetic research of many readers.

Kyoto, 10 June 2011 Hiroaki Ohno

On behalf of Yoshiji Takemoto
and Nobutaka Fujii

Acknowledgments

The author would like to express his sincere and wholehearted appreciation to Professor Nobutaka Fujii (Graduate School of Pharmaceutical Sciences, Kyoto University) for his kind guidance, constructive discussions and constant encouragement during this study. Thanks also go to Dr. Hiroaki Ohno (Graduate School of Pharmaceutical Sciences, Kyoto University), Dr. Shinya Oishi (Graduate School of Pharmaceutical Sciences, Kyoto University), and Professor Yoshiji Takemoto (Graduate School of Pharmaceutical Sciences, Kyoto University) for their valuable suggestions, guidance, and support throughout this study. The support and advice of Professor Akira Otaka (Institute of Health Biosciences: IHBS and Graduate School of Pharmaceutical Sciences, the University of Tokushima), Professor Hirokazu Tamamura (Institute of Biomaterials and Bioengineering, Tokyo Medical and Dental University), Dr. Satoshi Ueda and Dr. Tetsuo Narumi (Institute of Biomaterials and Bioengineering, Tokyo Medical and Dental University) were greatly appreciated.

The author would also like to acknowledge all his colleagues in the Department of Bioorganic Medicinal Chemistry, Graduate School of Pharmaceutical Sciences, Kyoto University, Mr. Michinori Tanaka, Mr. Kimio Hirano, Dr. Jérôme Cluzeau, Dr. Kazumi Kajiwara, Dr. Tsuyoshi Watabe, Dr. Fumihiko Katagiri, Dr. Ayumu Niida, Dr. Yoshikazu Sasaki, Mr. Hiroki Nishikawa, Dr. Kenji Tomita, Ms. Ai Esaka, Mr. Hiroaki Tanigaki, Ms. Mai Nomura, Dr. Yusuke Ohta, Dr. Akinori Okano, Mr. Teppei Ogawa, Dr. Toshiaki Watanabe, Ms. Saori Itoh, Ms. Yasuyo Kodera, Ms. Eriko Inokuchi, Mr. Manabu Katoh, Ms. Yukiko Goto, Mr. Toshijiro Miyatani, Mr. Kazuya Kobayashi, Mr. Yamato Suzuki, Mr. Ryo Masuda, Mr. Hirotaka Kamitani, Mr. Kentaro Watanabe, Mr. Zengye Hou, Mr. Hiroaki Chiba, Mr. Tsukasa Mizuhara, Mr. Yuji Yoshimitsu, Ms. Noriko Tanahara, Ms. Ryoko Hayashi, Mr. Akira Iwata, Mr. Shingo Mizushima for their valuable comments and assistance.

Finally, the author would like to thank the Japan Society for the Promotion of Science (JSPS) for financial support, Dr. Naoshige Akimoto for mass spectral measurements, and all the staff at the Elemental Analysis Center, Kyoto University.

Contents

Chapter 1
Introduction

Abstract Development of efficient synthetic approaches for biologically active compounds, including natural products, is a prominent goal of modern organic chemistry. Transition-metal-catalyzed domino/cascade reactions are a useful tool for the direct construction of complicated compounds. These reactions can enhance the synthetic efficiency, and minimize the requirement for separation processes and waste production (for reviews, see Refs. [1–5]). Allenes are an important class of compounds with unique reactivity because of their cumulative double bonds. They have hybrid characteristics of an alkene and an alkyne, which makes them highly reactive toward a wide range of transition metals. Therefore, many attractive reactions of allenic compounds by transition metal catalysis have been developed (for reviews, see Refs. [6–11]); palladium-catalyzed cyclizations of allenes and related compounds have been used extensively for construction of cyclic compounds (for recent books and reviews on palladium-catalyzed cyclization of allenes, see Refs. [12–15]).

Recently, the author's group reported that bromoallenes can function as synthetic equivalents of allyl dication in the presence of a palladium(0) catalyst and alcohol (Scheme 1.1, eq 1) [16, 17]. This reactivity is useful for the efficient introduction of two nucleophiles **2**, such as hydroxy, amine or carbon nucleophiles, into substrates **1** (eq 2). The mechanism for this reaction can be explained as follows (eq 3). Oxidative addition of bromoallenes **1** to palladium(0) produces allenylpalladium complex **4**, which undergoes transformation into η^3-propargylpalladium intermediates **5**. Subsequently, first nucleophilic attack at the central carbon atom of the palladium complexes **5**, followed by second nucleophilic substitution of the resulting η^3-allylpalladium complexes **6** produces the adducts **3**. The author's group expanded this methodology to the synthesis of medium-sized heterocycle **9** from bromoallene **7** (Scheme 1.2, eq 4) [16, 17]. More recently, the author's group also developed an intramolecular domino cyclization of bromoallenes such as **10**

S. Inuki, *Total Synthesis of Bioactive Natural Products by Palladium-Catalyzed Domino Cyclization of Allenes and Related Compounds*, Springer Theses,
DOI: 10.1007/978-4-431-54043-4_1, © Springer 2012

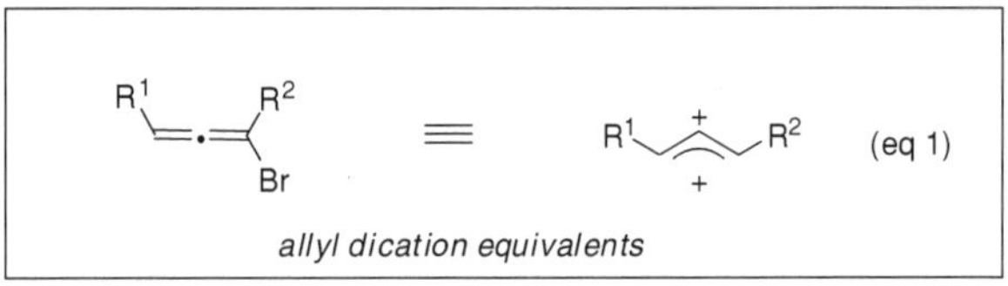

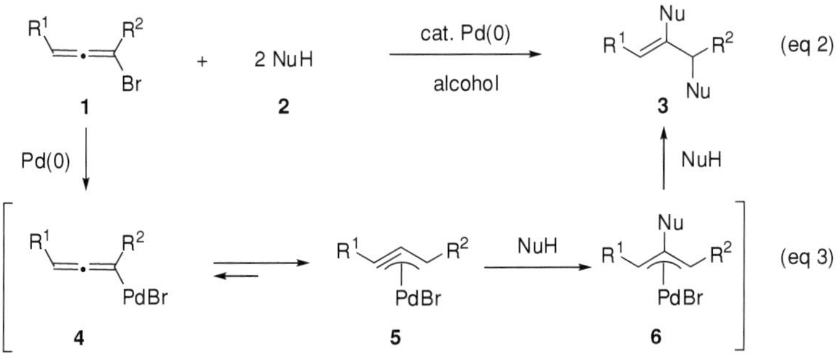

allyl dication equivalents

Scheme 1.1 Reaction of bromoallenes with two nucleophiles in the presence of a palladium(0) catalyst

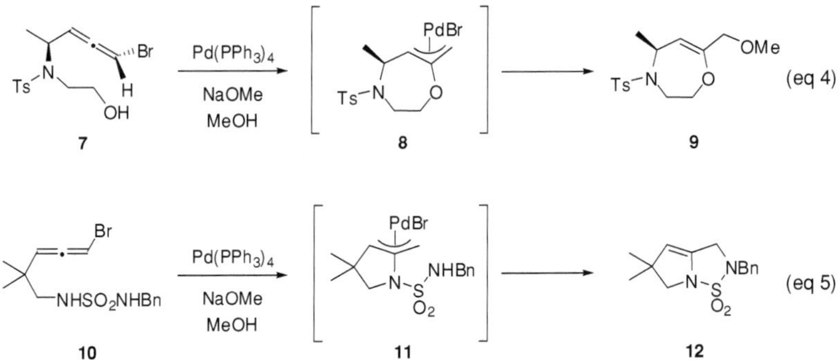

Scheme 1.2 Palladium(0)-catalyzed domino reactions by using bromoallenes as allyl dication equivalents

bearing a dual nucleophilic moiety, which produces bicyclic product **12** (Scheme 1.2, eq 5) [18, 19].

Meanwhile, the palladium-catalyzed reaction of propargylic compounds, developed by Tsuji et al., has become a useful tool for formation of two carbon–carbon or carbon–heteroatom bonds (Scheme 1.3, eq 6) (for pioneering works, see Refs. [20, 21]; for reviews on palladium-catalyzed reactions of propargylic compounds, see Refs. [21–23]; for representative examples of palladium-catalyzed reactions of propargylic compounds with nucleophiles, see Refs. [24–32]; for related reactions see Refs. [32–34]). Considerable research in this area has revealed that a combination of nucleophilic attacks by an internal nucleophilic

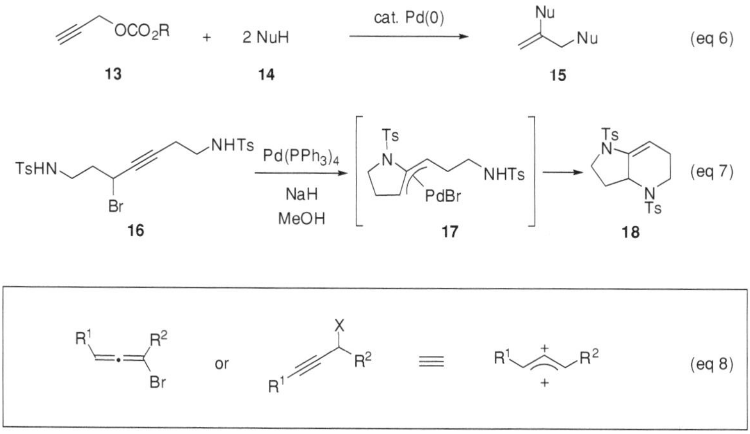

Scheme 1.3 Introduction of two nucleophiles into propargyl carbonates in the presence of a palladium(0) catalyst (Nu = nucleophile)

functional group and an appropriate external nucleophile can be an efficient approach to produce various cyclic compounds, such as carbapenems [28, 35–37], furans [38], indoles [39–41], indenes [42–44], and cyclic carbonates [45, 46]. Recently, the author's group developed a palladium(0)-catalyzed domino cyclization of propargyl bromides such as **16** with two nucleophilic sites, which produced bicyclic products **18** (Scheme 1.3, eq 7) [47, 48] (for a related work, see Ref. [49]). Thus, bromoallenes can be considered as a synthetic equivalent of propargylic compounds, and both can act as allyl dication equivalents (Scheme 1.3, eq 8).[1]

Based on these findings, domino cyclization of type **19** bromoallenes or type **20** propargyl compounds bearing nucleophilic groups at the both ends of a branched alkyl group was proposed, which would directly lead to bicyclic products **22** (Scheme 1.4). With this bis-cyclization as the key step, total synthesis of pachastrissamine, a biologically active natural product, was achieved.

The combination of aryl halides, allenes and nucleophiles such as amines and alcohols in the palladium(0)-catalyzed reaction enables the direct formation of carbon–carbon and carbon–heteroatom bonds (Scheme 1.5) (for related examples of palladium-catalyzed cyclization of amino allenes, see Refs. [50–67]; pioneering works on intermolecular Pd-catalyzed reactions of allenes, see Refs. [68, 69]; for related examples of palladium-catalyzed cyclization of amino allenes through the aminopalladation pathway, see Refs. [70–74]). This type of reaction may proceed through two pathways. The first pathway involves carbopalladation where the allenes are easily inserted into Pd(0)/R^3X derived aryl- or alkenylpalladium

[1] The reactivities of allenic and propargylic compounds are not necessarily the same. For example, propargyl bromides and carbonates are more reactive than bromoallenes toward S_N2 reactions and alcoholysis, respectively [17].

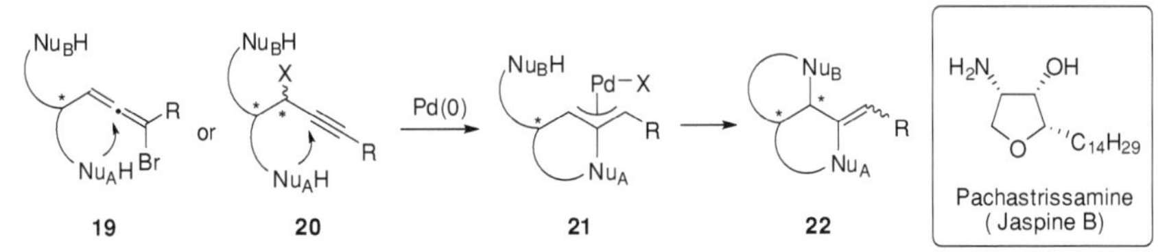

Scheme 1.4 Construction of bicyclic structures by palladium(0)-catalyzed cascade cyclization of bromoallenes **19** and propargyl compounds **20**

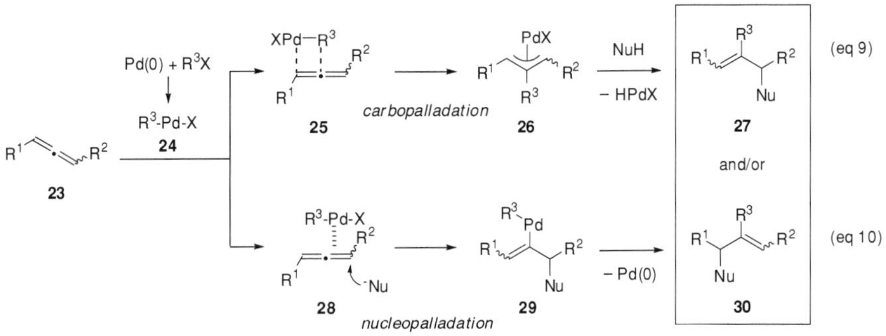

Scheme 1.5 Reaction of allenes with aryl- or alkenyl halides in the presence of a palladium(0) catalyst (Nu = nucleophile, X = halogen, R^3 = Aryl or Alkenyl)

halides **24**, which produces η^3-allylpalladium intermediates **26**. Subsequent nucleophilic attack of various nucleophiles on the η^3-allylpalladium intermediates produces allylic compounds **27** and/or **30** (Scheme 1.5, eq 9) [50–69, 75]. The second pathway involves nucleopalladation where the aryl- or alkenylpalladium halides coordinate to allenes, which promotes anti attack of the nucleophiles to give aryl- or alkenylpalladium complex **29**. This produces **27** and/or **30** by reductive elimination (Scheme 1.5, eq 10) [70–74, 76–80].

In 1984, Shimizu and Tsuji reported the first palladium-catalyzed intermolecular reaction of allene **31** with iodobenzene **32** and pyrrolidine **33** to afford 2,3-disubstituted allylic amine **34** (Scheme 1.6, eq 11) [68]. This methodology has since been extended to a wide range of heterocycle synthesis. Larock reported that palladium-catalyzed reaction of allene **36** with aryl halide derivatives such as 2-haloaniline **35**, bearing a nucleophilic functional group, directly produces benzene-fused heterocyclic compound **37** (eq 12) [69]. Gallagher reported palladium-catalyzed cyclization of amino allene **38** with iodobenzene **32** to produce pyrrolidine **39** (eq 13) [50]. Recently, the author's group developed palladium-catalyzed zipper-mode domino cyclization of allenic haloalkene **40** to afford fused bicyclic heterocycle **41** (eq 14) [63]. However, the reaction of allenes with an aryl halide and amino group at both ends of internal allenes is unprecedented.

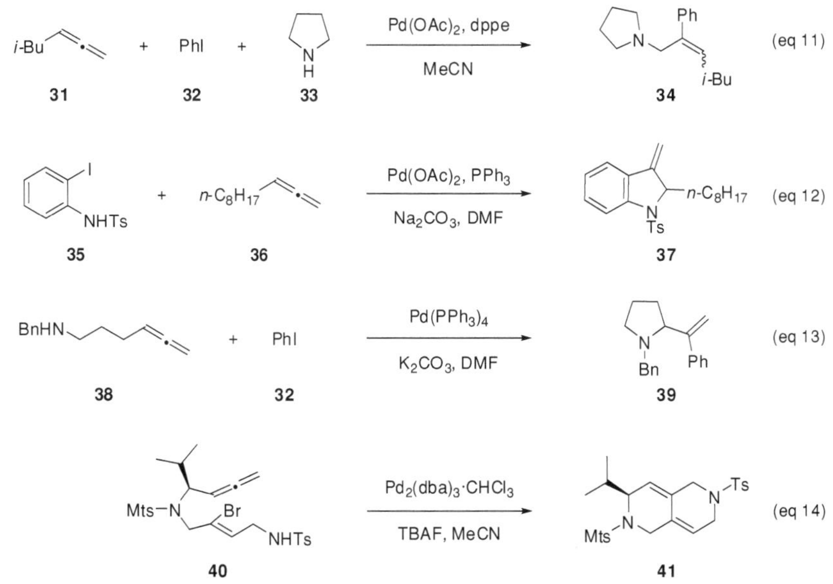

Scheme 1.6 Reactions of allenes with aryl- or alkenyl halides in the presence of a palladium(0) catalyst

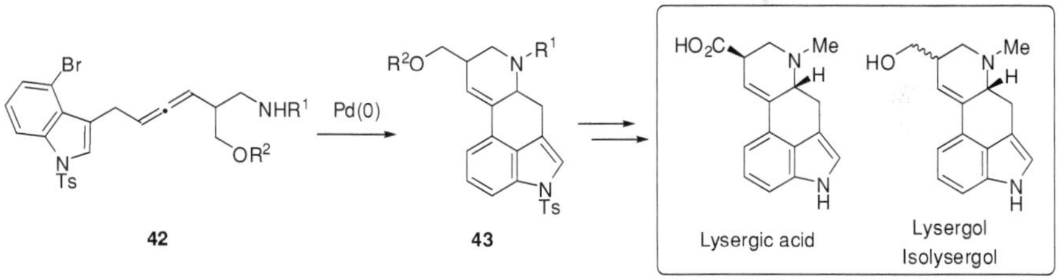

Scheme 1.7 Palladium(0)-catalyzed domino cyclization of amino allenes bearing a bromoindolyl group

In light of this chemistry, palladium-catalyzed domino cyclization of allene **42**, containing an appropriate nucleophilic group and aryl halide moiety, was proposed as a straightforward synthetic route for the core structure of ergot alkaloids **43** (Scheme 1.7). With this domino reaction as the key step, total synthesis of lysergic acid, lysergol and isolysergol was achieved.

In this study, total synthesis of the bioactive natural products, pachastrissamine, lysergic acid, lysergol and isolysergol, by palladium-catalyzed domino cyclization of allenes and related compounds was investigated.

Chapter 2 describes total synthesis of pachastrissamine (jaspine B) through palladium-catalyzed bis-cyclization of bromoallenes. This synthetic route was expanded to divergent synthesis of various pachastrissamine derivatives with different alkyl groups at the pachastrissamine C-2 position.

Chapter 3 describes total synthesis of pachastrissamine (jaspine B) through palladium-catalyzed bis-cyclization of propargyl chlorides and carbonates. This synthetic route furnished short-step synthesis of pachastrissamine in good overall yield (26% overall yield in 7 steps) from Garner's aldehyde.

Chapter 4 presents the total synthesis of (±)-lysergic acid, (±)-lysergol, and (±)-isolysergol by palladium-catalyzed domino cyclization of amino allenes bearing a bromoindolyl group, which was prepared via gold-catalyzed Claisen rearrangement.

Chapter 5 discusses enantioselective total synthesis of (+)-lysergic acid, (+)-lysergol, and (+)-isolysergol. The key intermediate, enantiomerically pure amino allene was prepared via palladium/indium-mediated reductive coupling reaction of L-serine-derived ethynylaziridine and Nozaki–Hiyama–Kishi (NHK) reaction. The synthesis highlights a strategy for constructing the C/D ring system of the core structure of ergot alkaloids based on palladium-catalyzed domino cyclization of amino allene, which allows creation of the stereochemistry at C5 by transfer of the axial chirality of allene to central chirality. This synthetic route furnished (+)-lysergic acid in 4.0% overall yield in 15 steps from the known ethynylaziridine.

References

1. Tietze LF (1996) Chem Rev 96:115–136
2. Tietze LF, Brasche G, Gericke K (2006) Domino reactions in organic synthesis. Wiley-VCH, Verlag GmbH, Weinheim
3. Nicolaou KC, Edmonds DJ, Bulger PG (2006) Angew Chem Int Ed 45:7134–7186
4. Padwa A, Bur SK (2007) Tetrahedron 63:5341–5378
5. Poulin J, Grisé-Bard CM, Barriault L (2009) Chem Soc Rev 38:3092–3101
6. Schuster H, Coppola G (1984) Allenes in organic synthesis. Wiley, New York
7. Pasto DJ (1984) Tetrahedron 40:2805–2827
8. Hashmi ASK (2000) Angew Chem Int Ed 39:3590–3593
9. Bates RW, Satcharoen V (2002) Chem Soc Rev 31:12–21
10. Ma S (2003) Acc Chem Res 36:701–712
11. Ma S (2005) Chem Rev 105:2829–2871
12. Yamamoto Y, Radhakrishnan U (1999) Chem Soc Rev 28:199–207
13. Zimmer R, Dinesh CU, Nandanan E, Khan FA (2000) Chem Rev 100:3067–3125
14. Mandai T (2004) In: Krause N, Hashmi ASK (eds) Modern allene chemistry, vol 2. Wiley-VCH, Weinheim, pp 925–972
15. Ohno H (2005) Chem Pharm Bull 53:1211–1226
16. Ohno H, Hamaguchi H, Ohata M, Tanaka T (2003) Angew Chem Int Ed 42:1749–1753
17. Ohno H, Hamaguchi H, Ohata M, Kosaka S, Tanaka T (2004) J Am Chem Soc 126:8744–8754
18. Hamaguchi H, Kosaka S, Ohno H, Tanaka T (2005) Angew Chem Int Ed 44:1513–1517
19. Hamaguchi H, Kosaka S, Ohno H, Fujii N, Tanaka T (2007) Chem Eur J 13:1692–1708
20. Tsuji J, Watanabe H, Minami I, Shimizu I (1985) J Am Chem Soc 107:2196–2198
21. Minami I, Yuhara M, Watanabe H, Tsuji J (1987) J Organomet Chem 334:225–242
22. Tsuji J, Minami I (1987) Acc Chem Res 20:140–145

23. Tsuji J, Mandai T (1995) Angew Chem Int Ed Engl 34:2589–2612
24. Minami I, Yuhara M, Tsuji J (1987) Tetrahedron Lett 28:629–632
25. Geng L, Lu X (1990) Tetrahedron Lett 31:111–114
26. Labrosse J-R, Lhoste P, Sinou D (1999) Tetrahedron Lett 40:9025–9028
27. Labrosse J-R, Lhoste P, Sinou D (2000) Org Lett 2:527–529
28. Labrosse J-R, Lhoste P, Sinou D (2001) J Org Chem 66:6634–6642
29. Zong K, Abboud KA, Reynolds JR (2004) Tetrahedron Lett 45:4973–4975
30. Yoshida M, Higuchi M, Shishido K (2008) Tetrahedron Lett 49:1678–1681
31. Yoshida M, Higuchi M, Shishido K (2009) Org Lett 11:4752–4755
32. Bi H-P, Liu X-Y, Gou F-R, Guo L-N, Duan X-H, Shu X-Z, Liang Y-M (2007) Angew Chem Int Ed 46:7068–7071
33. Ren Z-H, Guan Z-H, Liang Y-M (2009) J Org Chem 74:3145–3147
34. Gou F-R, Huo P-F, Bi H-P, Guan Z-H, Liang Y-M (2009) Org Lett 11:3418–3421
35. Kozawa Y, Mori M (2001) Tetrahedron Lett 42:4869–4873
36. Kozawa Y, Mori M (2002) Tetrahedron Lett 43:1499–1502
37. Kozawa Y, Mori M (2003) J Org Chem 68:8068–8074
38. Yoshida M, Morishita Y, Fujita M, Ihara M (2004) Tetrahedron Lett 45:1861–1864
39. Ambrogio I, Cacchi S, Fabrizi G (2006) Org Lett 8:2083–2086
40. Ambrogio I, Cacchi S, Fabrizi G, Prastaro A (2009) Tetrahedron 65:8916–8929
41. Cacchi S, Fabrizi G, Filisti, E (2009) Synlett 1817–1821
42. Duan X-H, Guo L-N, Bi H-P, Liu X-Y, Liang Y-M (2006) Org Lett 8:5777–5780
43. Guo L-N, Duan X-H, Bi H-P, Liu X-Y, Liang Y-M (2007) J Org Chem 72:1538–1540
44. Bi H-P, Guo L-N, Gou F-R, Duan X-H, Liu X-Y, Liang Y-M (2008) J Org Chem 73: 4713–4716
45. Yoshida M, Ihara M (2001) Angew Chem Int Ed 40:616–619
46. Yoshida M, Fujita M, Ishii T, Ihara M (2003) J Am Chem Soc 125:4874–4881
47. Ohno H, Okano A, Kosaka S, Tsukamoto K, Ohata M, Ishihara K, Maeda H, Tanaka T, Fujii N (2008) Org Lett 10:1171–1174
48. Okano A, Tsukamoto K, Kosaka S, Maeda H, Oishi S, Tanaka T, Fujii N, Ohno H (2010) Chem Eur J 16:8410–8418
49. Okano A, Oishi S, Tanaka T, Fujii N, Ohno H (2010) J Org Chem 75:3396–3400
50. Davies IW, Scopes DIC, Gallagher T (1993) Synlett 85–87
51. Kang S-K, Baik T-G, Kulak AN (1999) Synlett 324–326
52. Rutjes FPJT, Tjen KCMF, Wolf LB, Karstens WFJ, Schoemaker HE, Hiemstra H (1999) Org Lett 1:717–720
53. Ohno H, Toda A, Miwa Y, Taga T, Osawa E, Yamaoka Y, Fujii N, Ibuka T (1999) J Org Chem 64:2992–2993
54. Kang S-K, Baik T-G, Hur Y (1999) Tetrahedron 55:6863–6870
55. Anzai M, Toda A, Ohno H, Takemoto Y, Fujii N, Ibuka T (1999) Tetrahedron Lett 40: 7393–7397
56. Kang S-K, Kim K-J (2001) Org Lett 3:511–514
57. Hiroi K, Hiratsuka Y, Watanabe K, Abe I, Kato F, Hiroi M (2001) Synlett 263–265
58. Ohno H, Anzai M, Toda A, Oishi S, Fujii N, Tanaka T, Takemoto Y, Ibuka T (2001) J Org Chem 66:4904–4914
59. Grigg R, Köppen I, Rasparini M, Sridharan V (2001) Chem Commun 964–965
60. Hiroi K, Hiratsuka Y, Watanabe K, Abe I, Kato F, Hiroi M (2002) Tetrahedron Asymm 13:1351–1353
61. Watanabe K, Hiroi K (2003) Heterocycles 59:453–457
62. Grigg R, Inman M, Kilner C, Köppen I, Marchbank J, Selby P, Sridharan V (2007) Tetrahedron 63:6152–6169
63. Okano A, Mizutani T, Oishi S, Tanaka T, Ohno H, Fujii N (2008) Chem Commun 3534–3536
64. Cheng X, Ma S (2008) Angew Chem Int Ed 47:4581–4583
65. Beccalli EM, Broggini G, Clerici F, Galli S, Kammerer C, Rigamonti M, Sottocornola S (2009) Org Lett 11:1563–1566

66. Shu W, Ma S (2009) Chem Commun 6198–6200
67. Beccalli EM, Bernasconi A, Borsini E, Broggini G, Rigamonti M, Zecchi G (2010) J Org Chem 75:6923–6932
68. Shimizu I, Tsuji J (1984) Chem Lett 233–236
69. Larock RC, Berrios-Peña NG, Fried CA (1991) J Org Chem 56:2615–2617
70. Karstens WFJ, Rutjes FPJT, Hiemstra H (1997) Tetrahedron Lett 38:6275–6278
71. Karstens WFJ, Stol M, Rutjes FPJT, Hiemstra H (1998) Synlett 1126–1128
72. Ma S, Gao W (2002) Org Lett 4:2989–2992
73. Ma S, Yu F, Gao W (2003) J Org Chem 68:5943–5949
74. Ma S, Yu F, Li J, Gao W (2007) Chem Eur J 13:247–254
75. Stevens RR, Shier GD (1970) J Organometal Chem 21:495–499
76. Lathbury D, Vernon P, Gallagher T (1986) Tetrahedron Lett 27:6009–6012
77. Prasad JS, Liebeskind LS (1988) Tetrahedron Lett 29:4257–4260
78. Fox DNA, Lathbury D, Mahon MF, Molloy KC, Gallagher T (1991) J Am Chem Soc 113:2652–2656
79. Kimura M, Fugami K, Tanaka S, Tamaru Y (1992) J Org Chem 57:6377–6379
80. Kimura M, Tanaka S, Tamaru Y (1995) J Org Chem 60:3764–3772

Part I
Total Synthesis of Pachastrissamine (Jaspine B)

Chapter 2
Total Synthesis through Palladium-Catalyzed Bis-Cyclization of Bromoallenes

Abstract Palladium(0)-catalyzed cyclization of bromoallenes bearing hydroxy and benzamide groups as internal nucleophiles stereoselectively provides functionalized tetrahydrofuran. This cyclization was expanded to divergent synthesis of pachastrissamine, a biologically active marine natural product, and its derivatives.

Pachastrissamine **1** (Fig. 2.1), an anhydrophytosphingosine derivative isolated from a marine sponge *Pachastrissa* sp., was reported by Higa et al. in 2002 [1]. Shortly thereafter, Debitus et al. isolated the same compound from a different marine sponge, *Jaspis* sp., and named jaspine B [2]. Other structurally related analogues have also been isolated from the same species, including jaspine A and 2-*epi*-jaspine B. Pachastrissamine (jaspine B) **1** exhibits cytotoxic activity against various tumor cell lines at nanomolar level [1, 2]. In 2009, Delgado et al. reported that dihydroceramides mediated autophagy might be involved in the cytotoxicity [3]. Andrieu-Abadie et al. indicated that pachastrissamine induces apoptotic cell death in melanoma cells by a caspase-dependent pathway [4]. Owing to its biological importance, pachastrissamine has been the target of many synthetic studies (for previous syntheses [5–23]). Stereoselective construction of the tetrahydrofuran ring which bears three contiguous stereogenic centers is a major issue in the total synthesis.

As described in Chap. 1, the author planned domino cyclization of type **2** bromoallenes or type **3** propargyl compounds bearing nucleophilic groups at the both ends of a branched alkyl group, which would directly lead to bicyclic products such as **6** or **7** (Scheme 2.1). This bis-cyclization also enables a cyclization/functionalization cascade, which creates a new chiral center on the *exo*-type second cyclization and utilizes the chiral center at the branched position. The key to success of this domino reaction would be controlled successive nucleophilic attacks by Nu_A and Nu_B in the desired order. First cyclization by Nu_A or Nu_B will produce intermediate **4** or **5**, respectively. These would be converted to the cyclic products **6** or **7**, respectively, by the second intramolecular reaction.

S. Inuki, *Total Synthesis of Bioactive Natural Products by Palladium-Catalyzed Domino Cyclization of Allenes and Related Compounds*, Springer Theses,
DOI: 10.1007/978-4-431-54043-4_2, © Springer 2012

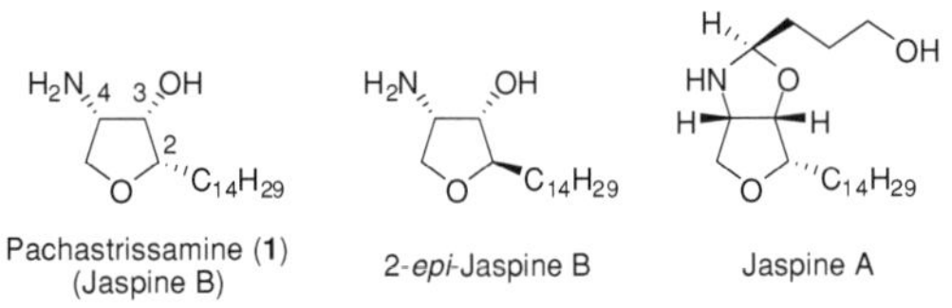

Fig. 2.1 Structures of naturally occurring jaspines

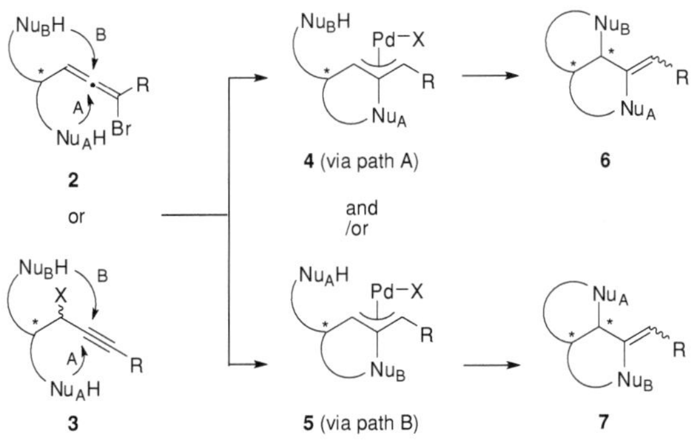

Scheme 2.1 Construction of bicyclic structures by palladium(0)-catalyzed cascade cyclization of bromoallenes **2** and propargyl compounds **3**

The author is also interested in the stereochemical course of the domino cyclization, *i.e.* the effect of the axial or central chirality in the allenic/propargylic moiety of **2**/**3** on the reactivity and selectivity. The author chose pachastrissamine **1** for the model study to evaluate this working hypothesis on the ring-construction/ stereoselective functionalization cascade.

The author expected that palladium(0)-catalyzed cyclization of bromoallenes **9** bearing hydroxy and benzamide groups [24–26] as internal nucleophiles could regio- and stereoselectively provide appropriately functionalized tetrahydrofuran **8** for synthesis of pachastrissamine **1** (Scheme 2.2). The bicyclic structure of **8** including the exo-olefin would be useful for stereoselective construction of a C-2 stereogenic center as well as carbon homologation. This synthetic route takes an advantage of the late-stage introduction of the long alkyl side chain into the tetrahydrofuran ring at the C-2 position, which makes it possible to achieve a divergent synthesis of pachastrissamine derivatives.

Preparation of the required bromoallene **9a** is outlined in Scheme 2.3. The *erythro*-alkynol **11a** was easily prepared from (*S*)-Garner's aldehyde **10** [27, 28] following the literature procedure [29]. Treatment of **11a** with MsCl and Et_3N gave the corresponding mesylate, which was then allowed to react with CuBr·DMS/LiBr [30, 31] (DMS = Me_2S) to afford the (*S*,a*R*)-bromoallene **12a.**

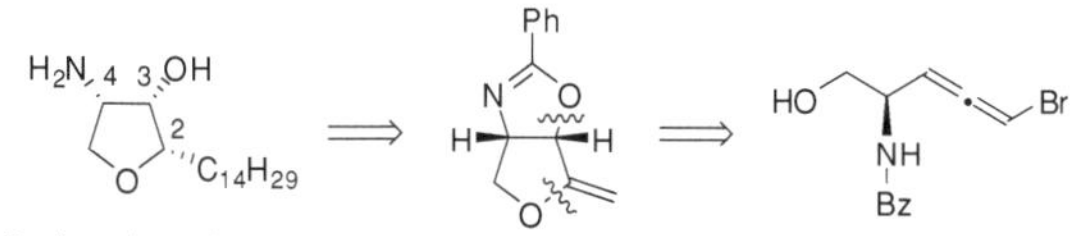

Scheme 2.2 Retrosynthetic analysis of pachastrissamine 1

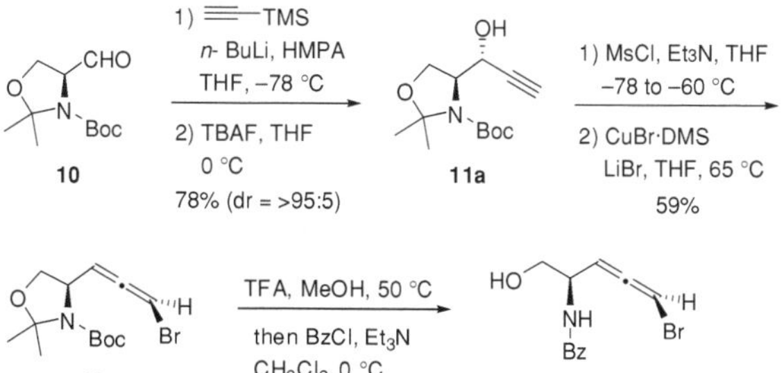

Scheme 2.3 Synthesis of bromoallene 9a

(preparation of 12 was previously reported in Refs. [32, 33]).[1] Removal of the Boc and acetal groups with TFA followed by acylation with BzCl/Et_3N afforded the benzamide **9a**.

The author next investigated cascade cyclization of bromoallene **9a** in the presence of palladium(0) (Table 2.1). Treatment of **9a** with $Pd(PPh_3)_4$ (5 mol %) and NaH (2.0 equiv) in MeOH at 50 °C (standard conditions for cyclization of bromoallenes) [34, 35] successfully produced the desired bicyclic tetrahydrofuran **8** in 50% yield (entry 1). The undesired cyclization initiated by the first cyclization by the benzamide group (Scheme 2.1) was not promoted. However, the anticipated side-products dihydrofuran **13a** (formed by the intermolecular reaction with methoxide) and a small amount of furan **14** were observed. Formation of the furan **14** can be rationalized by β-hydride elimination of the η^3-allylpalladium intermediate (e.g. **4** or **5**, Scheme 2.1) followed by aromatization. (a related furan formation as a by-product in the cascade cyclization of propargylic bromides was recently reported, see Ref. [36]). To suppress the intermolecular reaction with the external alkoxide, the reaction was examined under other conditions, including the use of a mixed solvent. Reaction in THF/MeOH (4:1) decreased yields of both **8** and **13a** (40 and 15%, respectively), while the amount of furan **14** increased (10% yield, entry 2). Of the several bases investigated, Cs_2CO_3 (1.2 equiv) most effectively produced the desired product **8** and suppressed formation of furan **14** (entries 2–5). The best result was obtained using a mixed solvent of THF/MeOH

[1] For improvement of the yield of **12**, a slightly modified bromination protocol was used (3 equiv of the copper reagent, 65 °C; see the Experimental Section).

Table 2.1 Palladium-catalyzed cascade cyclization of bromoallene **9a**[a]

9a → $Pd(PPh_3)_4$, base, solvent, 50 °C → **8** + **13** + **14**

13a: R = CH_3
13b: R = CH_2CF_3

Entry	Base (equiv)	Solvent	Yield (%)[b]			Recovery (%)[c]
			8	**13**	**14**	
1	NaH (2.0)	MeOH	50	45	trace	–
2	NaH (2.0)	THF/MeOH (4:1)	40	15	10	–
3	K_2CO_3 (2.0)	THF/MeOH (4:1)	43	–	–	41
4	Cs_2CO_3 (2.0)	THF/MeOH (4:1)	67	26	–	–
5	Cs_2CO_3 (1.2)	THF/MeOH (4:1)	78	20	–	–
6	Cs_2CO_3 (1.2)	THF/MeOH (10:1)	89	trace	–	–
7	Cs_2CO_3 (1.2)	THF	12	–	–	64
8	Cs_2CO_3 (2.0)	THF/TFE (4:1)	–	93	–	–
9	Cs_2CO_3 (2.0)	THF/*t*-BuOH (4:1)	12	–	–	60

[a] All reactions were performed with 5 mol % $Pd(PPh_3)_4$ at 0.1 M for 1–4 h
[b] Yield of isolated products
[c] Recovery of starting material. TFE = 2,2,2-trifluoroethanol

(10:1) in the presence of 1.2 equiv of Cs_2CO_3 (89%, entry 6). It should be noted that the use of solely THF resulted in low yield of **8** (12%, entry 7) and recovery of the starting material, which suggests that an alcoholic solvent plays an important role in this type of transformation. Interestingly, use of CF_3CH_2OH, a more acidic solvent which might facilitate the protonation step, only gave the undesired compound **13b** bearing a trifluoroethoxy group in high yield (93%, entry 8). Moreover, use of *t*-BuOH was not effective (entry 9). These results indicate that pK_a values and bulkiness of the alcoholic solvents have significant effects on the reaction, i.e. the intramolecular vs. intermolecular reaction in the second nucleophilic attack, and reactivity of the bromoallene with a palladium catalyst.

To investigate the difference in reactivity between the diastereomeric bromoallenes **9a** and **9b**, the author next synthesized (*S*,a*S*)-bromoallene **9b**, also starting from Garner's aldehyde **10** (Scheme 2.4). The *threo*-alkynol **11b**, stereoselectively obtained following Taddei's protocol (preparation of **12** was previously reported in Ref. [32, 33])[2], was converted into the desired bromoallene **9b** in the same manner as described above (Scheme 2.3). Bromoallene **9b** was then subjected to the optimized reaction conditions shown in entry 6 (Table 2.1) to give the desired bicyclic product **8** in 88% yield. These results show both bromoallenes **9a** and **9b** equally undergo the cascade cyclization to give the same product **8**. This means that a diastereomeric mixture of bromoallenes can be directly employed for preparation of **8**.

[2] For improvement of the yield of **12**, a slightly modified bromination protocol was used (3 equiv of the copper reagent, 65 °C; see the Experimental Section).

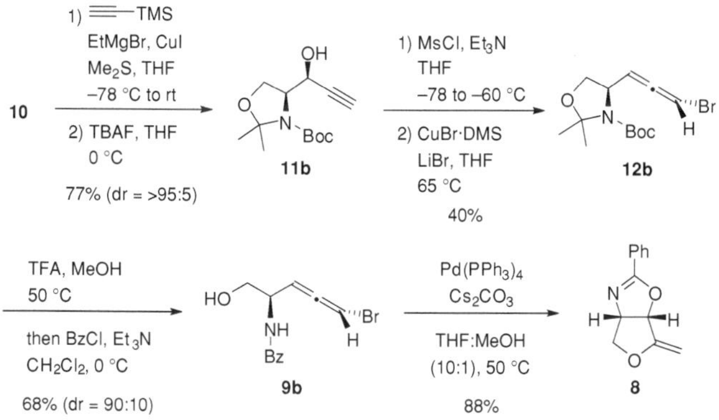

Scheme 2.4 Synthesis and palladium-catalyzed cascade cyclization of the epimeric bromoallene **9b**

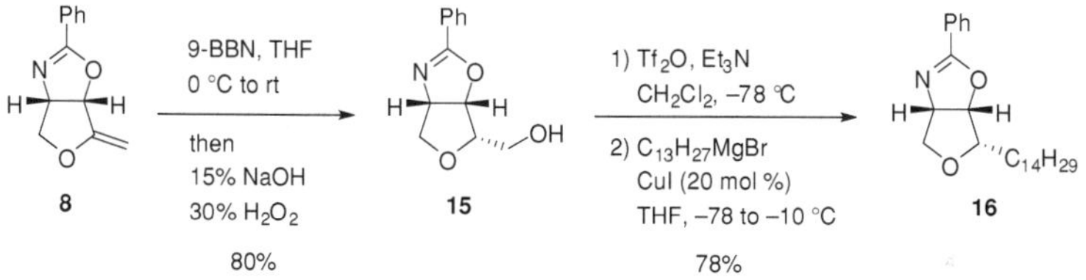

Scheme 2.5 Synthesis of protected pachastrissamine (**16**)

With the functionalized tetrahydrofuran **8** prepared, the author examined the introduction of a C-2 alkyl side chain with an all-*cis* configuration. Hydroboration-oxidation of the exo-olefin of **8** with 9-BBN provided the primary alcohol **15** with the desired configuration as the sole diastereomer [37]. Treatment of **15** with Tf_2O and Et_3N followed by displacement with a cuprate derived from $C_{13}H_{27}MgBr/CuI$ provided the tetrahydrofuran **16** bearing all the requisite functionalities (Scheme 2.5) [38]. The cleavage of oxazoline ring will be described in Chap. 3.

The author next investigated the incorporation of diverse side chains into the C-2 position using a variety of organocopper reagents derived from Grignard reagents (Table 2.2). Reaction with Grignard reagents containing a primary alkyl group such as phenylethyl and methyl in the presence of a copper salt (20 mol %) afforded the desired alkylation products in good yields (entries 1, 2) [39, 40]. Changing the Grignard reagents to *i*-PrMgCl or $CH_2{=}C(CH_3)MgBr$ gave moderate yields of the corresponding products **17c** or **17d** (entries 3, 4), respectively, containing a secondary alkyl or alkenyl group [41]. The author next examined introduction of an allyl group, which can be readily used for further manipulation (Table 2.3). However, treatment of the triflate with allylMgBr and catalytic CuBr [39, 40] provided the unanticipated product oxaazabicycloheptane **18** in 90% yield (entry 1). (Structure of **18** was confirmed by NMR analysis and comparison with structurally related compounds, see Ref. [42]). The reaction in the absence of a copper catalyst, also afforded **18** in 91% yield (entry 2). In comparison, use of

Table 2.2 Copper-catalyzed alkylation of triflates[a]

1) Tf_2O, Et_3N, CH_2Cl_2, −78 °C; 2) RMgX, Cu cat. (20 mol %), solvent, temp.
15 → **17**

Entry	RMgX	CuX cat.	Solvent	Temp.	Products (% yield)[b]
1	$Ph(CH_2)_2MgCl$	CuI	THF	−78 °C to rt	**17a** (76)
2	MeMgBr	CuBr	$THF:Et_2O$ (9:1)	−30 °C to rt	**17b** (82)
3	*i*-PrMgCl	$CuBr{\cdot}Me_2S$	$THF:Me_2S$ (30:1)	−20 to 0 °C	**17c** (54)
4	$CH_2{=}C(CH_3)MgBr$	$CuBr{\cdot}Me_2S$	$THF:Me_2S$ (30:1)	−20 to 0 °C	**17d** (66)

[a] Reactions were carried out with RMgX (2.7–7.0 equiv) and CuX (20 mol%) for 1–4.5 h
[b] Isolated yields

Table 2.3 Copper-catalyzed allylation of triflates and formation of 2-oxa-5-azabicyclo [2.2.1]heptanes

1) Tf_2O, Et_3N, CH_2Cl_2, −78 °C; 2) conditions
15 → **17e** + **18**

Entry	Conditions	Yield (%)[a,b]	
		17e	**18**
1[c]	AllylMgBr, CuBr (20 mol%), $THF:Et_2O$ (3:1), −30 °C	ND	90
2[c]	AllylMgBr, $THF:Et_2O$ (3:1), −30 °C	ND	91
3[d]	$(Allyl)_2Cu(CN)Li_2$, THF, −78 °C	32	ca. 5

[a] Isolated yields
[b] ND = Not detected
[c] Reactions were carried out with allylMgBr (5.0 equiv) for 1.5 h
[d] Reactions were carried out with $(allyl)_2Cu(CN)Li_2$ (4.0 equiv) for 30 min

$(allyl)_2Cu(CN)Li_2$ [43, 44] resulted in 32% yield of the desired product **17e** along with a small amount of the side product **18** (entry 3).

Rationale for formation of the oxaazabicycloheptane **18** is depicted in Scheme 2.6. The addition of allylMgBr to imine followed by intramolecular attack of the resulting nitrogen anion to triflate would generate mono-allylated intermediate **21**. The second nucleophilic attack of allylMgBr to iminium cation **22** derived from **21** would proceed to give the oxaazabicycloheptane **18**.[3] Servi et al. also reported that 2-phenyloxazolines bearing a tosylate leaving group with allyl Grignard reagent gave bicyclic compounds similar to the intermediate **21** [45]. In contrast, the reaction with the other Grignard reagents did not afford the

[3] Reaction of triflate **19** with 1.0 equiv of the allyl Grignard reagents gave the oxaazabicycloheptane **18** in 13% yield along with the recovery of the unchanged triflate **19** in 73% yield, without isolation of the intermediate **21**. This is presumably due to a highly strained aminal structure of **21** and facile Grignard reaction to the iminium moiety of **22**.

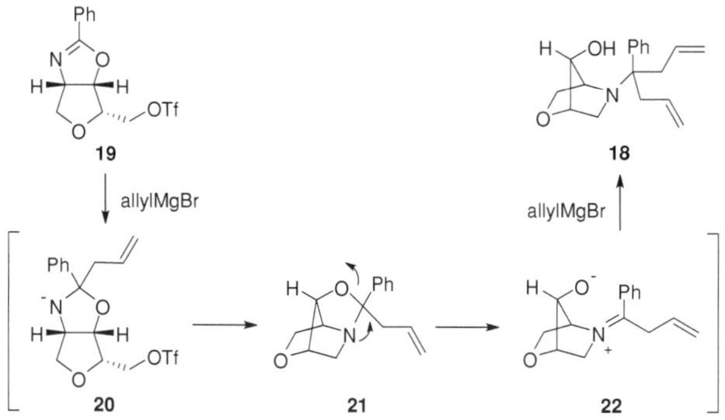

Scheme 2.6 Formation of 2-oxa-5-azabicyclo[2.2.1]heptane **18**

oxaazabicycloheptane-type products (Table 2.2). The formation of the oxaazabicycloheptane **18** with allylMgBr would be caused by the first addition to imine **19** proceeding through six-membered transition state.

In conclusion, the author has developed a novel ring-construction/stereoselective functionalization cascade by palladium(0)-catalyzed bis-cyclization of bromoallenes. Using bromoallenes bearing hydroxy and benzamide groups as internal nucleophiles allows the sequential nucleophilic reactions to selectively proceed in the desired order to form a functionalized tetrahydrofuran ring. This strategy provides an efficient synthetic route to protected pachastrissamine **16** and its derivatives **17** bearing three contiguous stereogenic centers from Garner's aldehyde as the sole chiral source.

2.1 Experimental Section

2.1.1 General Methods

All moisture-sensitive reaction were performed using syringe-septum cap techniques under an argon atmosphere and all glassware was dried in an oven at 80 °C for 2 h prior to use. Reactions at −78 °C employed a CO_2–MeOH bath. Melting points were measured by a hot stage melting point apparatus (uncorrected). Optical rotations were measured with a JASCO P-1020 polarimeter. For flash chromatography, Wakosil C-300, Wakogel C-300E or Chromatorex® was employed. ^{1}H NMR spectra were recorded using a JEOR AL-400 or JEOL ECA-500 spectrometer, and chemical shifts are reported in δ (ppm) relative to TMS (in $CDCl_3$) as internal standard. ^{13}C NMR spectra were recorded using a JEOL AL-400 or JEOL ECA-500 spectrometer and referenced to the residual $CHCl_3$ signal. ^{19}F NMR spectra were recorded using a JEOL ECA-500 and referenced to the internal $CFCl_3$ (δ_F 0.00 ppm). ^{1}H NMR spectra are tabulated as follows: chemical shift, multiplicity (b = broad, s = singlet, d = doublet, t = triplet, q = quartet, m = multiplet), number of protons, and coupling constant(s). Exact mass (HRMS) spectra were recorded on a JMS-HX/HX 110A mass spectrometer. Infrared (IR) spectra were obtained on a JASCO FT/IR-4100 FT-IR spectrometer with JASCO ATR PRO410-S.

2.1.2 *tert-Butyl (R)-4-[(R)-3-Bromopropa-1,2-dienyl]-2,2-dimethyloxazolidine-3-carboxylate (12a)*

To a stirred mixture of the propargylic alcohol **11a** [29] (5.66 g, 22.2 mmol) and Et_3N (15.4 mL, 111 mmol) in THF (70 mL) was added MsCl (3.40 mL, 44.4 mmol) at −78 °C, and the mixture was stirred for 0.5 h with warming to −60 °C. The mixture was made acidic with saturated NH_4Cl at −60 °C, and the mixture was concentrated under reduced pressure. The residue was extracted with Et_2O. The extract was washed with H_2O and brine and was dried over Na_2SO_4. Concentration of the filtrate under reduced pressure followed by rapid filtration through a short pad of silica gel with Et_2O to give a crude mesylate, which was used without further purification. A mixture of CuBr·DMS (13.7 g, 66.6 mmol) and LiBr (5.80 g, 66.6 mmol) was dissolved in THF (70 mL) at room temperature under argon. After stirring for 2 min, a solution of the above crude mesylate in THF (90 mL) was added to this reagent at room temperature. The mixture was stirred at 65 °C for 4 h and quenched with saturated NH_4Cl and 28% NH_4OH. The whole was extracted with Et_2O and the extract was washed with H_2O and brine and was dried over Na_2SO_4. Concentration under reduced pressure followed by flash chromatography over silica gel with *n*-hexane–EtOAc (13:1) to give **12a** as a colorless oil (4.15 g, 59% yield). All spectral data were in agreement with those reported by Taddei [33].

2.1.3 *N-[(2R,4R)-5-Bromo-1-hydroxypenta-3,4-dien-2-yl] benzamide (9a)*

To a stirred solution of **12a** (200 mg, 0.629 mmol) in MeOH (0.30 mL) at 0 °C was added trifluoroacetic acid (1 mL), and the mixture was stirred at 50 °C for 1 h. The mixture was concentrated under reduced pressure, and the residue was dissolved in CH_2Cl_2 (5 mL). The solution was made neutral with Et_3N at 0 °C. Further Et_3N (0.306 mL, 2.20 mmol) and BzCl (0.080 mL, 0.692 mmol) were added to the stirred mixture at 0 °C. The mixture was stirred at this temperature for 4.5 h, followed by quenching with H_2O. The whole was extracted with EtOAc. The extract was washed successively with 1 N HCl, H_2O and brine, and dried over Na_2SO_4. The filtrate was concentrated under reduced pressure to give an oily residue, which was purified by flash chromatography over silica gel with *n*-hexane–EtOAc (2:3) to give **9a** as a white solid (98.7 mg, 56% yield). Recrystallization from *n*-hexane–EtOAc gave pure **9a** as colorless crystals: mp 149–150 °C; $[\alpha]_D^{25}$ −240.7 (*c* 1.22, MeOH); IR (neat): 3340 (OH), 1963 (C=C=C), 1627 (C=O); ^1H NMR (500 MHz, $CDCl_3$) δ 2.35 (t, J = 6.0 Hz, 1H), 3.85–3.93 (m, 2H), 4.92–4.99 (m, 1H), 5.61 (dd, J = 5.7, 4.6 Hz, 1H), 6.20 (dd, J = 5.7, 2.9 Hz, 1H), 6.58 (d, J = 7.4 Hz, 1H), 7.46 (dd, J = 7.4,

7.4 Hz, 2H), 7.54 (t, $J = 7.4$ Hz, 1H), 7.80 (d, $J = 7.4$ Hz, 2H); ^{13}C NMR (125 MHz, DMSO) δ 50.3, 62.7, 74.9, 100.7, 127.3 (2C), 128.2 (2C), 131.2, 134.4, 166.1, 200.8. *Anal.* Calcd for $C_{12}H_{12}BrNO_2$: C, 51.09; H, 4.29; N, 4.96. Found: C, 51.18; H, 4.22; N, 5.00.

2.1.4 (3aS,6aS)-6-Methylene-2-phenyl-3a,4,6,6a -tetrahydrofuro[3,4-d]oxazole (8)

To a stirred mixture of **9a** (40 mg, 0.142 mmol) in THF/MeOH (1.2 mL, 10:1) were added $Pd(PPh_3)_4$ (8.2 mg, 0.0071 mmol) and Cs_2CO_3 (55.5 mg, 0.170 mmol) at room temperature under argon (Table 2.1, Entry 6). The mixture was stirred at 50 °C for 2.5 h, and filtered through a short pad of silica gel with EtOAc to give a crude **8**. The filtrate was concentrated under reduced pressure to give a yellow oil, which was purified by flash chromatography over silica gel with *n*-hexane–EtOAc (3:1) to give **8** as a white solid (25.5 mg, 89% yield): mp 98–99 °C; $[\alpha]_D^{25}+287.4$ (*c* 1.05, $CHCl_3$); IR (neat): 1641 (C=O); ^{1}H NMR (500 MHz, $CDCl_3$) δ 4.32–4.37 (m, 2H), 4.43 (dd, $J = 2.0$, 0.7 Hz, 1H), 4.59–4.62 (m, 1H), 4.92 (ddd, $J = 8.0$, 5.4, 2.9 Hz, 1H), 5.42 (d, $J = 8.0$ Hz, 1H), 7.42 (dd, $J = 6.8$, 6.8 Hz, 2H), 7.50 (tt, $J = 6.8$, 1.7 Hz, 1H), 7.94 (d, $J = 6.8$ Hz, 2H); ^{13}C NMR (125 MHz, $CDCl_3$) δ 70.2, 76.2, 81.6, 87.5, 127.0, 128.4 (2C), 128.5 (2C), 131.7, 161.3, 164.2. *Anal.* Calcd for $C_{12}H_{11}NO_2$: C, 71.63; H, 5.51; N, 6.96. Found: C, 71.43; H, 5.70; N, 6.89.

2.1.5 (R)-N-[5-(Methoxymethyl)-2,3-dihydrofuran-3-yl] benzamide (13a)

Yellow oil; $[\alpha]_D^{26}-82.7$ (*c* 1.77, $CHCl_3$); IR (neat): 1634 (C=O); ^{1}H NMR (500 MHz, $CDCl_3$) δ 3.42 (s, 3H), 4.00 (d, $J = 13.2$ Hz, 1H), 4.03 (d, $J = 13.2$ Hz, 1H), 4.32 (dd, $J = 10.3$, 3.2 Hz, 1H), 4.58 (dd, $J = 10.3$, 8.6 Hz, 1H), 5.09 (d, $J = 2.3$ Hz, 1H), 5.28–5.35 (m, 1H), 6.32 (d, $J = 6.3$ Hz, 1H), 7.42 (dd, $J = 7.4$, 7.4 Hz, 2H), 7.50 (t, $J = 7.4$ Hz, 1H), 7.75 (d, $J = 7.4$ Hz, 2H); ^{13}C NMR (125 MHz, $CDCl_3$) δ 53.2, 59.0, 67.1, 77.4, 97.8, 126.9 (2C), 128.5 (2C), 131.6, 133.9, 159.9, 166.8; HRMS (FAB) calcd $C_{13}H_{16}NO_3$: $[M+H]^+$, 234.1130; found: 234.1130.

2.1.6 (R)-N-{5-[(2,2,2-Trifluoroethoxy)methyl] -2,3-dihydrofuran-3-yl}benzamide (13b)

Pale yellow solid; mp 100–101 °C; $[\alpha]_D^{27}-87.9$ (*c* 1.16, $CHCl_3$); IR (neat): 1629 (C=O); ^{1}H NMR (500 MHz, $CDCl_3$) δ 3.91 (q, $J_{C\text{–}F} = 8.6$ Hz, 2H), 4.22 (s, 2H), 4.32 (dd, $J = 10.3$, 3.2 Hz, 1H), 4.58 (d, $J = 10.3$, 8.6 Hz, 1H), 5.14 (d,

$J = 2.9$ Hz, 1H), 5.29–5.36 (m, 1H), 6.38 (d, $J = 6.9$ Hz, 1H), 7.42 (dd, $J = 7.4$, 7.4 Hz, 2H), 7.51 (t, $J = 7.4$ Hz, 1H), 7.76 (d, $J = 7.4$ Hz, 2H); ^{13}C NMR (125 MHz, $CDCl_3$) δ 53.1, 66.7, 68.1 (q, $J_{C-F} = 34.8$ Hz), 77.5, 98.9, 123.8 (q, $J_{C-F} = 278.3$ Hz), 126.9 (2C), 128.6 (2C), 131.7, 133.8, 158.5, 166.9; ^{19}F NMR (471 MHz, $CFCl_3$) δ –74.0 (3F). *Anal.* Calcd for $C_{14}H_{14}F_3NO_3$: C, 55.82; H, 4.68; N, 4.65. Found: C, 56.05; H, 4.82; N, 4.50.

2.1.7 N-(5-Methylfuran-3-yl)benzamide (14)

Yellow solid; mp 128–130 °C; IR (neat): 1643 (C=O); ^{1}H NMR (500 MHz, $CDCl_3$) δ 2.29 (s, 3H), 6.04 (s, 1H), 7.42 (dd, $J = 7.7$, 7.7 Hz, 2H), 7.54 (t, $J = 7.7$ Hz, 1H), 7.67 (s, 1H), 7.83 (d, $J = 7.7$ Hz, 2H), 8.00 (s, 1H); ^{13}C NMR (125 MHz, $CDCl_3$) δ 13.6, 101.0, 124.7, 126.9 (2C), 128.7 (2C), 131.0, 131.8, 134.0, 151.1, 164.8; HRMS (FAB) calcd $C_{12}H_{12}NO_2$: $[M+H]^+$, 202.0868; found: 202.0864.

2.1.8 tert-Butyl (R)-4-[(S)-3-Bromopropa-1,2-dienyl]-2,2-dimethyloxazolidine-3-carboxylate (12b)

By a procedure identical with that described for synthesis of **12a** from **11a**, the propargylic alcohol **11b** (1.82 g, 7.13 mmol) was converted into **12b** as a colorless oil (902 mg, 40% yield): $[\alpha]_D^{25}$ +34.5 (*c* 1.30, $CHCl_3$); IR (neat): 1962 (C=C=C), 1697 (C=O); ^{1}H NMR (500 MHz, $CDCl_3$, 50 °C) δ 1.49 (s, 9H), 1.51 (s, 3H), 1.60 (s, 3H), 3.89 (dd, $J = 8.9$, 1.1 Hz, 1H), 4.06 (d, $J = 8.9$, 6.0 Hz, 1H), 4.36–4.65 (m, 1H), 5.40–5.55 (m, 1H), 6.05–6.11 (m, 1H); ^{13}C NMR (125 MHz, $CDCl_3$, 50 °C) δ 23.7 (0.5C), 24.9 (0.5C), 26.5 (0.5C), 27.2 (0.5C), 28.5 (3C), 55.5, 67.8, 74.3, 80.4, 94.4, 101.0, 151.8, 201.8; HRMS (FAB) calcd $C_{13}H_{21}BrNO_3$: $[M+H]^+$, 318.0705; found: 318.0708.

2.1.9 N-[(2R,4S)-5-Bromo-1-hydroxypenta-3,4-dien-2-yl]benzamide (9b)

By a procedure identical with that described for synthesis of **9a** from **12a**, the bromoallene **12b** (841 mg, 2.64 mmol) was converted into **9b** as a white solid (510 mg, 68% yield, dr = 10:1). Recrystallization from *n*-hexane–EtOAc gave **9b** (dr = 90:10) as colorless crystals: mp 110–111 °C; $[\alpha]_D^{26}$ +248.4 (*c* 1.18, MeOH, dr = 90:10); IR (neat): 3323 (OH), 1960 (C=C=C), 1639 (C=O); ^{1}H NMR (500 MHz, $CDCl_3$) δ 2.44 (t, $J = 6.0$ Hz, 1H), 3.89 (dd, $J = 5.4$, 4.3 Hz, 2H), 4.89–4.97 (m, 1H), 5.60 (dd, $J = 5.7$, 4.9 Hz, 1H), 6.22 (dd, $J = 5.7$, 2.9 Hz, 1H), 6.62 (d, $J = 6.3$ Hz, 1H), 7.45 (dd, $J = 7.4$, 7.4 Hz, 2H), 7.53 (t, $J = 7.4$ Hz, 1H), 7.80 (d, $J = 7.4$ Hz, 2H); ^{13}C NMR (125 MHz, $CDCl_3$) δ 49.9, 64.6, 75.6, 99.4,

127.1 (2C), 128.7 (2C), 131.9, 133.9, 167.8, 201.5. *Anal.* Calcd for $C_{12}H_{12}BrNO_2$: C, 51.09; H, 4.29; N, 4.96. Found: C, 50.89; H, 4.37; N, 4.69.

2.1.10 [(3aS,6S,6aS)-2-Phenyl-3a,4,6,6a-tetrahydrofuro [3,4-d]oxazol-6-yl]methanol (15)

To a stirred mixture of **8** (352 mg, 1.75 mmol) in THF (7 mL) were added 9-BBN (0.5 M solution in THF; 10.5 mL, 5.25 mmol) at 0 °C under argon. After stirring at this temperature for 30 min and at room temperature for additional 10 min, the mixture was cooled to 0 °C and quenched by the careful addition of 15% NaOH (5 mL) and 30% H_2O_2 (5 mL). The mixture was stirred at room temperature for 1.5 h, followed by quenching with saturated NH_4Cl. The whole was extracted with Et_2O. The extract was washed with brine, and dried over Na_2SO_4. The filtrate was concentrated under reduced pressure to give an oily residue, which was purified by flash chromatography over silica gel with *n*-hexane–EtOAc (1:6) to give **15** as a white solid (309 mg, 80% yield). Recrystallization from *n*-hexane–EtOAc gave pure **15** as colorless crystals: mp 130–131 °C; $[\alpha]_D^{24}$ +51.4 (*c* 1.03, $CHCl_3$); IR (neat): 3363 (OH), 1648 (C=O); ^{1}H NMR (400 MHz, $CDCl_3$) δ 1.91 (dd, J = 7.3, 4.9 Hz, 1H), 3.81 (dd, J = 10.0, 5.4 Hz, 1H), 3.85–3.90 (m, 1H), 3.90–4.03 (m, 2H), 4.19 (d, J = 10.0 Hz, 1H), 4.92 (dd, J = 7.7, 5.4 Hz, 1H), 5.13 (dd, J = 7.7, 3.8 Hz, 1H), 7.42 (dd, J = 7.3, 7.3 Hz, 2H), 7.50 (t, J = 7.3 Hz, 1H), 7.91 (d, J = 7.3 Hz, 2H); ^{13}C NMR (125 MHz, $CDCl_3$) δ 61.0, 72.8, 73.5, 82.8, 83.9, 126.8, 128.3 (2C), 128.4 (2C), 131.6, 164.3. *Anal.* Calcd for $C_{12}H_{13}NO_3$: C, 65.74; H, 5.98; N, 6.39. Found: C, 65.81; H, 5.85; N, 6.38.

2.1.11 (3aS,6S,6aS)-2-Phenyl-6-tetradecyl-3a,4,6, 6a-tetrahydrofuro[3,4-d]oxazole (16)

To a stirred mixture of **15** (735 mg, 3.35 mmol) and Et_3N (0.93 mL, 6.70 mmol) in CH_2Cl_2 (33 mL) was added Tf_2O (0.79 mL, 4.69 mmol) at −78 °C, and the mixture was stirred for 30 min. The mixture was quenched by addition of saturated NH_4Cl at −78 °C, and the whole was extracted with CH_2Cl_2. The extract was washed with H_2O and brine and was dried over Na_2SO_4. Concentration of the filtrate under reduced pressure followed by rapid filtration through a short pad of silica gel with Et_2O–CH_2Cl_2 (1:1) to give a crude triflate, which was used without further purification. To a suspension of CuI (128 mg, 0.67 mmol) in THF (15 mL) was added dropwise a solution of $C_{13}H_{27}MgBr$ in THF (0.75 M; 12.1 mL, 9.05 mmol) at −78 °C under argon. The mixture was allowed to warm to 0 °C, and was stirred at this temperature for 10 min. To the mixture of the resulting cuprate was added dropwise a solution of the above triflate in THF

(28 mL) at −78 °C, and the mixture was allowed to warm to −10 °C. After stirring at this temperature for 30 min, the mixture was quenched by addition of saturated NH_4Cl and 28% NH_4OH. The whole was extracted with Et_2O and the extract was washed with H_2O and dried over Na_2SO_4. Concentration under reduced pressure followed by flash chromatography over Chromatorex® with *n*-hexane–EtOAc (6:1) gave **16** as a white solid (1.00 g, 78% yield): mp 93–94 °C; $[\alpha]_D^{24}$+60.9 (*c* 1.05, $CHCl_3$); IR (neat): 1651 (C=O); ^{1}H NMR (500 MHz, $CDCl_3$) δ 0.88 (t, J = 6.9 Hz, 3H), 1.21–1.40 (m, 22H), 1.43–1.56 (m, 2H), 1.76 (dt, J = 7.0, 7.0 Hz, 2H), 3.62 (td, J = 7.0, 4.0 Hz, 1H), 3.72 (dd, J = 10.3, 5.4 Hz, 1H), 4.11 (d, J = 10.3 Hz, 1H), 4.83 (dd, J = 7.7, 5.4 Hz, 1H), 4.99 (dd, J = 7.7, 4.0 Hz, 1H), 7.40 (dd, J = 7.4, 7.4 Hz, 2H), 7.48 (t, J = 7.4 Hz, 1H), 7.93 (d, J = 7.4 Hz, 2H); ^{13}C NMR (125 MHz, $CDCl_3$) δ 14.1, 22.7, 26.1, 28.7, 29.3, 29.5, 29.6 (6C), 29.7, 31.9, 72.6, 73.2, 83.6, 84.0, 127.3, 128.3 (4C) 131.4, 164.3. *Anal.* Calcd for $C_{25}H_{39}NO_2$: C, 77.87; H, 10.19; N, 3.63. Found: C, 77.76; H, 10.42; N, 3.51.

2.1.12 (3aS,6S,6aS)-2-Phenyl-6-(3-phenylpropyl)-3a,4,6,6a-tetrahydrofuro[3,4-d]oxazole (17a)

By a procedure identical with that described for synthesis of **16** from **15**, the alcohol **15** (40 mg, 0.18 mmol) was converted into the corresponding crude triflate, which was used without further purification (Table 2.2, Entry 1). To a suspension of CuI (6.9 mg, 0.036 mmol) in THF (0.9 mL) was added dropwise a solution of $Ph(CH_2)_2MgCl$ in THF (1.0 M; 0.90 mL, 0.90 mmol) at −78 °C under argon. The mixture was allowed to warm to 0 °C, and was stirred for 10 min at this temperature. To the mixture of the resulting cuprate was added dropwise a solution of the above triflate in THF (1.3 mL) at −78 °C, and the mixture was allowed to warm to room temperature. After stirring for 1.5 h at this temperature, the mixture was quenched by addition of saturated NH_4Cl and 28% NH_4OH. The whole was extracted with Et_2O and the extract was washed with H_2O and brine, and was dried over Na_2SO_4. Concentration under reduced pressure followed by flash chromatography over silica gel with *n*-hexane–EtOAc (1:1) and then over Chromatorex® with *n*-hexane–EtOAc (2:1) gave **17a** as a colorless oil (42.2 mg, 76% yield): $[\alpha]_D^{26}$+97.7 (*c* 1.49, $CHCl_3$); IR (neat): 1650 (C=N); ^{1}H NMR (500 MHz, $CDCl_3$) δ 1.77–1.83 (m, 2H), 1.81–1.89 (m, 2H), 2.63–2.70 (m, 1H), 2.71–2.77 (m, 1H), 3.60–3.64 (m, 1H), 3.70 (dd, J = 9.7, 5.4 Hz, 1H), 4.09 (d, J = 9.7 Hz, 1H), 4.81 (dd, J = 7.7, 5.4 Hz, 1H), 4.96 (dd, J = 7.7, 3.7 Hz, 1H), 7.17–7.23 (m, 3H), 7.28 (dd, J = 7.4, 7.4 Hz, 2H), 7.38 (dd, J = 7.7, 7.7 Hz, 2H), 7.46 (t, J = 7.7 Hz, 1H), 7.85 (d, J = 7.7 Hz, 2H); ^{13}C NMR (125 MHz, $CDCl_3$) δ 27.8, 28.4, 35.9, 72.7, 73.2, 83.5, 83.8, 125.8, 127.2, 128.3 (4C), 128.5 (4C), 131.4, 142.1, 164.3; HRMS (FAB) calcd for $C_{20}H_{22}NO_2$ $[M+H]^+$, 308.1651, found: 308.1655.

2.1.13 (3aS,6S,6aS)-6-Ethyl-2-phenyl-3a,4,6,6a-tetrahydrofuro[3,4-d]oxazole (17b)

By a procedure identical with that described for synthesis of **16** from **15**, the alcohol **15** (40 mg, 0.18 mmol) was converted into the corresponding crude triflate, which was used without further purification (Table 2.2, Entry 2). To a suspension of CuBr (5.2 mg, 0.036 mmol) in THF (1.6 mL) was added dropwise a solution of MeMgBr in Et_2O (3.0 M; 0.30 mL, 0.90 mmol) at 0 °C under argon. The mixture was stirred for 10 min at this temperature. To the mixture of the resulting cuprate was added dropwise a solution of the above triflate in THF (1.1 mL) at −30 °C. After stirring for 1.5 h at this temperature, the mixture was allowed to warm to room temperature. The mixture was stirred for 3.0 h at room temperature and quenched with saturated NH_4Cl and 28% NH_4OH. The whole was extracted with Et_2O and the extract was washed with H_2O and brine, and was dried over Na_2SO_4. Concentration under reduced pressure followed by flash chromatography over silica gel with *n*-hexane–EtOAc (2:3) gave **17b** as a white waxy solid (32.0 mg, 82% yield): mp 56–57 °C; $[\alpha]_D^{26}$ +79.7 (*c* 1.02, $CHCl_3$); IR (neat): 1650 (C=N); ^1H NMR (500 MHz, $CDCl_3$) δ 1.09 (t, J = 7.4 Hz, 3H), 1.74–1.86 (m, 2H), 3.56 (ddd, J = 6.9, 6.9, 3.4 Hz, 1H), 3.72 (dd, J = 9.7, 5.4 Hz, 1H), 4.11 (d, J = 9.7 Hz, 1H), 4.84 (dd, J = 7.4, 5.4 Hz, 1H), 5.00 (dd, J = 7.4, 3.4 Hz, 1H), 7.40 (dd, J = 7.4, 7.4 Hz, 2H), 7.47 (t, J = 7.4 Hz, 1H), 7.93 (d, J = 7.4 Hz, 2H); ^{13}C NMR (125 MHz, $CDCl_3$) δ 10.4, 22.0, 72.6, 73.2, 83.3, 85.3, 127.3, 128.3 (4C), 131.4, 164.3; *Anal.* Calcd for $C_{13}H_{15}NO_2$: C, 71.87; H, 6.96; N, 6.45. Found: C, 71.58; H, 7.05; N, 6.38.

2.1.14 (3aS,6S,6aS)-6-Isobutyl-2-phenyl-3a,4,6,6a-tetrahydrofuro[3,4-d]oxazole (17c)

By a procedure identical with that described for synthesis of **16** from **15**, the alcohol **15** (40 mg, 0.18 mmol) was converted into the corresponding crude triflate, which was used without further purification (Table 2.2, Entry 3). To a solution of the above triflate and CuBr·Me_2S (7.4 mg, 0.036 mmol) in THF/Me_2S (2.1 mL, 20:1) was added dropwise a solution of *i*-PrMgCl in THF (1.5 M; 0.84 mL, 1.26 mmol) at −20 °C under argon. After stirring for 2.0 h at this temperature, the mixture was allowed to warm to 0 °C. The mixture was stirred for 1.0 h at this temperature and quenched with saturated NH_4Cl and 28% NH_4OH. The whole was extracted with Et_2O and the extract was washed with H_2O and brine, and was dried over Na_2SO_4. Concentration under reduced pressure followed by flash chromatography over silica gel with *n*-hexane–EtOAc (2:3) and then over Chromatorex® with *n*-hexane–EtOAc (3:1) gave **17c** as a white solid (23.8 mg, 54% yield). Recrystallization from *n*-hexane–EtOAc gave pure **17c** as colorless crystals: mp 79–80 °C; $[\alpha]_D^{26}$+71.5 (*c* 0.96, $CHCl_3$); IR (neat): 1650 (C=N); ^{1}H NMR (500 MHz, $CDCl_3$) δ 0.99 (d, J = 6.3 Hz, 3H), 1.00 (d, J = 6.3 Hz, 3H), 1.62 (ddd, J = 13.7, 7.4, 6.7 Hz, 1H), 1.69 (ddd, J = 13.7, 7.4, 6.9 Hz, 1H), 1.82–1.90 (m, 1H), 3.68–3.71 (m, 1H), 3.71 (dd, J = 9.7, 5.4 Hz, 1H), 4.11 (d, J = 9.7 Hz, 1H), 4.83 (dd, J = 7.4, 5.4 Hz, 1H), 4.98 (dd, J = 7.4, 4.0 Hz, 1H), 7.40 (dd, J = 7.4, 7.4 Hz, 2H), 7.47 (t, J = 7.4 Hz, 1H), 7.93 (d, J = 7.4 Hz, 2H); ^{13}C NMR (125 MHz, $CDCl_3$) δ 22.6, 23.2, 25.3, 37.6, 72.7, 73.2, 82.3, 84.0, 127.3, 128.3 (4C), 131.4, 164.3. *Anal.* Calcd for $C_{15}H_{19}NO_2$: C, 73.44; H, 7.81; N, 5.71. Found: C, 73.19; H, 7.81; N, 5.60.

2.1.15 (3aS,6S,6aS)-6-(2-Methylallyl)-2-phenyl -3a,4,6,6a-tetrahydrofuro[3,4-d]oxazole (17d)

By a procedure identical with that described for synthesis of **16** from **15**, the alcohol **15** (40 mg, 0.18 mmol) was converted into the corresponding crude triflate, which was used without further purification (Table 2.2, Entry 4). To a solution of the above triflate and $CuBr{\cdot}Me_2S$ (7.4 mg, 0.036 mmol) in THF/Me_2S (1.0 mL, 9:1) was added dropwise a solution of CH_2=C(CH_3)MgBr in THF (0.5 M; 1.8 mL, 0.90 mmol) at −20 °C under argon. After stirring for 2.0 h at this temperature, the mixture was allowed to warm to 0 °C. The mixture was stirred for 1.0 h at this temperature and quenched with saturated NH_4Cl and 28% NH_4OH. The whole was extracted with Et_2O and the extract was washed with H_2O and brine, and was dried over Na_2SO_4. Concentration under reduced pressure followed by flash chromatography over silica gel with *n*-hexane–EtOAc (2:3) and then over Chromatorex® with *n*-hexane–EtOAc (3:1) gave **17d** as a white solid (28.8 mg, 66% yield): mp 55–56 °C; $[\alpha]_D^{26}$+85.6 (*c* 1.10, $CHCl_3$); IR (neat): 1650 (C=N); ^{1}H NMR (500 MHz, $CDCl_3$) δ 1.85 (s, 3H), 2.47 (dd, J = 14.9, 7.4 Hz, 1H), 2.52 (dd, J = 14.9, 6.3 Hz, 1H), 3.74 (dd, J = 9.7, 5.4 Hz, 1H), 3.81 (ddd, J = 7.4, 6.3, 3.7 Hz, 1H), 4.13 (d, J = 9.7 Hz, 1H), 4.84 (dd, J = 7.7, 5.4 Hz, 1H), 4.89–4.91 (m, 2H), 5.01 (dd, J = 7.7, 3.7 Hz, 1H), 7.41 (dd, J = 7.4, 7.4 Hz, 2H), 7.48 (t, J = 7.4 Hz, 1H), 7.94 (d, J = 7.4 Hz, 2H); ^{13}C NMR (125 MHz, $CDCl_3$) δ 22.9, 36.8, 72.6, 73.4, 82.2, 83.6, 112.6, 127.2, 128.3 (4C), 131.4, 141.9, 164.3. *Anal.* Calcd for $C_{15}H_{17}NO_2$: C, 74.05; H, 7.04; N, 5.76. Found: C, 73.77; H, 7.09; N, 5.64.

2.1.16 (1S,4S,7S)-5-(4-Phenylhepta-1,6-dien-4-yl) -2-oxa-5-azabicyclo[2.2.1]heptan-7-ol (18)

By a procedure identical with that described for synthesis of **16** from **15**, the alcohol **15** (40 mg, 0.18 mmol) was converted into the corresponding crude triflate, which was used without further purification (Table 2.3, Entry 2). To a mixture of allylMgBr in Et_2O (1.0 M; 0.90 mL, 0.90 mmol) in THF (1.6 mL) was added dropwise a solution of the above triflate in THF (1.1 mL) at −30 °C under argon. After stirring for 1.5 h at this temperature, the mixture was quenched with saturated NH_4Cl. The whole was extracted with Et_2O and the extract was washed with H_2O and brine, and was dried over Na_2SO_4. Concentration under reduced pressure followed by flash chromatography over silica gel with *n*-hexane–EtOAc (3:2) gave **18** as a colorless oil (46.5 mg, 91% yield): $[\alpha]_D^{25}$ +35.9 (*c* 1.61, $CHCl_3$); IR (neat): 3445 (OH); ^1H NMR (500 MHz, $CDCl_3$) δ 2.79 (dd, J = 14.6, 8.6 Hz, 2H), 2.85 (dd, J = 14.6, 8.3 Hz, 2H), 2.97 (d, J = 9.5 Hz, 1H), 2.90–2.99 (m, 1H), 3.13 (d, J = 8.0 Hz, 1H), 3.16 (dd, J = 9.5, 2.3 Hz, 1H), 3.38–3.40 (m, 1H), 3.52 (dd, J = 8.0, 1.7 Hz, 1H), 3.96 (dd, J = 2.3, 2.3 Hz, 1H), 4.02–4.04 (m, 1H), 5.09 (d, J = 10.3 Hz, 1H), 5.12 (d, J = 10.3 Hz, 1H), 5.17 (d, J = 16.6 Hz, 2H), 5.68–5.79 (m, 2H), 7.28 (t, J = 7.4 Hz, 1H), 7.36 (dd, J = 7.4, 7.4 Hz, 2H), 7.46 (d, J = 7.4 Hz, 2H); ^{13}C NMR (125 MHz, $CDCl_3$) δ 39.3, 39.6, 49.0, 59.0, 60.4, 70.5, 73.8, 77.9, 118.3, 118.5, 126.6 (2C), 127.3, 128.5 (2C), 133.8 (2C), 141.7; HRMS (FAB) calcd for $C_{18}H_{24}NO_2$: $[M+H]^+$, 286.1807, found: 286.1805.

2.1.17 (3aS,6S,6aS)-6-(But-3-enyl)-2-phenyl -3a,4,6,6a-tetrahydrofuro[3,4-d]oxazole (17e)

By a procedure identical with that described for synthesis of **16** from **15**, the alcohol **15** (40 mg, 0.18 mmol) was converted into the corresponding crude triflate, which was used without further purification (Table 2.3, Entry 3). To a suspension of CuCN (71.6 mg, 0.72 mmol) in THF (2.0 mL) was added dropwise a solution of MeLi in Et_2O (1.06 M; 1.36 mL, 1.44 mmol) at −78 °C under argon. The mixture was allowed to warm to 0 °C, and was stirred for 10 min at this temperature. To the mixture was added dropwise allyltributylstannane (0.45 mL, 1.44 mmol) at −78 °C, and the mixture was allowed to warm to room temperature. The mixture was stirred for 30 min at this temperature. To the mixture of the resulting cuprate was added dropwise a solution of the above triflate in THF (1.1 mL) at −78 °C. After stirring for 30 min at this temperature, the mixture was quenched with saturated NH_4Cl and 28% NH_4OH. The whole was extracted with Et_2O and the extract was washed with H_2O and brine, and was dried over Na_2SO_4. Concentration under reduced pressure followed by flash chromatography over silica gel with *n*-hexane–EtOAc (10:1 to 2:3) gave **17e** as a white waxy solid (14.0 mg, 32% yield): mp 55–56 °C; $[\alpha]_D^{24}$ +91.7 (*c* 0.50, $CHCl_3$); IR (neat): 1651 (C=N); ^1H NMR (500 MHz, $CDCl_3$) δ 1.82–1.94 (m, 2H), 2.22–2.35 (m, 2H), 3.65 (ddd, J = 6.9, 6.9, 3.4 Hz, 1H), 3.72 (dd, J = 9.7, 5.4 Hz, 1H), 4.12 (d, J = 9.7 Hz, 1H), 4.84 (dd, J = 7.7, 5.4 Hz, 1H), 5.00 (dd, J = 7.7, 3.4 Hz, 1H), 5.02 (d, J = 10.3 Hz, 1H), 5.10 (d, J = 16.6 Hz, 1H), 5.88 (ddd, J = 16.6, 10.3, 6.8 Hz, 1H), 7.41 (dd, J = 7.7, 7.7 Hz, 2H), 7.48 (t, J = 7.7 Hz, 1H), 7.93 (d, J = 7.7 Hz, 2H); ^{13}C NMR (125 MHz, $CDCl_3$) δ 28.0, 30.3, 72.8, 73.2, 83.2,

83.5, 115.1, 127.3, 128.3 (4C), 131.4, 138.0, 164.3; HRMS (FAB) calcd for $C_{15}H_{18}NO_2$ $[M+H]^+$, 244.1338, found: 244.1338.

References

1. Kuroda I, Musman M, Ohtani I, Ichiba T, Tanaka J, Garcia-Gravalos D, Higa T (2002) J Nat Prod 65:1505–1506
2. Ledroit V, Debitus C, Lavaud C, Massoit G (2003) Tetrahedron Lett 44:225–228
3. Canals D, Mormeneo D, Fabriàs G, Llebaria A, Casas J, Delgado A (2009) Bioorg Med Chem 17:235–241
4. Salma Y, Lafort E, Therville N, Carpentier S, Bonnafé M-J, Levade T, Génisson Y, Andrieu-Abadie N (2009) Biochem Pharmacol 78:477–485
5. Sudhakar N, Kumar AR, Prabhakar A, Jagadeesh B, Rao BV (2005) Tetrahedron Lett 46:325–327
6. Bhaket P, Morris K, Stauffer CS, Datta A (2005) Org Lett 7:875–876
7. van den Berg R, Boltje T, Verhagen C, Litjens R, Vander Marel G, Overkleeft H (2006) J Org Chem 71:836–839
8. Du Y, Liu J, Linhardt RJ (2006) J Org Chem 71:1251–1253
9. Liu J, Du Y, Dong X, Meng S, Xiao J, Cheng L (2006) Carbohydr Res 341:2653–2657
10. Ribes C, Falomir E, Carda M, Marco JA (2006) Tetrahedron 62:5421–5425
11. Lee T, Lee S, Kwak YS, Kim D, Kim S (2007) Org Lett 9:429–432
12. Reddy LVR, Reddy PV, Shaw AK (2007) Tetrahedron Asymmetr 18:542–546
13. Ramana CV, Giri AG, Suryawanshi SB, Gonnade RG (2007) Tetrahedron Lett 48:265–268
14. Prasad KR, Chandrakumar A (2007) J Org Chem 72:6312–6315
15. Abraham E, Candela-Lena JI, Davies SG, Georgiou M, Nicholson RL, Roberts PM, Russell AJ, Snchez-Fernndez EM, Smith AD, Thomson JE (2007) Tetrahedron Asymmetr 18:2510–2513
16. Yakura T, Sato S, Yoshimoto Y (2007) Chem Pharm Bull 55:1284–1286
17. Abraham E, Brock EA, Candela-Lena JI, Davies SG, Georgiou M, Nicholson RL, Perkins JH, Roberts PM, Russell AJ, Snchez-Fernndez EM, Scott PM, Smith AD, Thomson JE (2008) Org Biomol Chem 6:1665–1673
18. Passiniemi M, Koskinen AMP (2008) Tetrahedron Lett 49:980–983
19. Venkatesan K, Srinivasan KV (2008) Tetrahedron Asymmetr 19:209–215
20. Enders D, Terteryan V, Palecek J (2008) Synthesis 2278–2282
21. Ichikawa Y, Matsunaga K, Masuda T, Kotsuki H, Nakano K (2008) Tetrahedron 64:11313–11318
22. Yoshimitsu Y, Inuki S, Oishi S, Fujii N, Ohno H (2010) J Org Chem 75:3843–3846
23. Abraham E, Davies SG, Roberts PM, Russell AJ, Thomson JE (2008) Tetrahedron: Asymmetry 19:1027–1047
24. Cook GR, Shanker PS (1998) Tetrahedron Lett 39:3405–3408
25. Cook GR, Shanker PS (1998) Tetrahedron Lett 39:4991–4994
26. Lee K-Y, Kim Y-H, Park M-S, Oh C-Y, Ham W-H (1999) J Org Chem 64:9450–9458
27. Garner P (1984) Tetrahedron Lett 25:5855–5858
28. Campbell AD, Raynham TM, Taylor RJK (1998) Synthesis 1707–1709
29. Herold P (1988) Helv Chim Acta 71:354–362
30. Montury M, Goré J (1980) Synth Commun 10:873–879
31. Elsevier CJ, Meijer J, Tadema G, Stehouwer PM, Bos HJT, Vermeer P (1982) J Org Chem 47:2194–2196
32. D'Aniello F, Mann A, Taddei M, Wermuth C-G (1994) Tetrahedron Lett 35:7775–7778
33. D'Aniello F, Mann A, Schoenfelder A, Taddei M (1997) Tetrahedron 53:1447–1456
34. Ohno H, Hamaguchi H, Ohata M, Tanaka T (2003) Angew Chem Int Ed 42:1749–1753

35. Ohno H, Hamaguchi H, Ohata M, Kosaka S, Tanaka T (2004) J Am Chem Soc 126:8744–8754
36. Ohno H, Okano A, Kosaka S, Tsukamoto K, Ohata M, Ishihara K, Maeda H, Tanaka T, Fujii N (2008) Org Lett 10:1171–1174
37. Ghosh AK, Xi K (2007) Org Lett 9:4013–4016
38. Evans PA, Cui J, Gharpure SJ, Polosukhin A, Zhang H-R (2003) J Am Chem Soc 125:14702–14703
39. Kotsuki H, Kadota I, Ochi M (1990) J Org Chem 55:4417–4422
40. Somfai P (1994) Tetrahedron 50:11315–11320
41. Arnold LD, Drover JCG, Vedreras JC (1987) J Am Chem Soc 109:4649–4659
42. Hümmer W, Dubois E, Gracza T, Jäger V (1997) Synthesis 634–642
43. Lipshutz BH, Crow R, Dimock SH, Ellsworth ELJ (1990) J Am Chem Soc 112:4063–4064
44. Lipshutz BH, Ellsworth EL, Dimock SH, Smith RAKJ (1990) J Am Chem Soc 112:4404–4410
45. Fronza G, Mele A, Pedrocchi-Fantoni G, Pizzi D, Servi S (1990) J Org Chem 55:6216–6219

Chapter 3
Total Synthesis through Palladium-Catalyzed Bis-Cyclization of Propargyl Chlorides and Carbonates

Abstract Palladium(0)-catalyzed cyclization of propargyl chlorides and carbonates bearing hydroxy and benzamide groups as internal nucleophiles stereoselectively provides functionalized tetrahydrofuran. Cyclization reactivity is dependent on the relative configuration of the benzamide and leaving groups, and on the nature of the leaving groups. This bis-cyclization was used as the key step in a short-step total synthesis of pachastrissamine.

Based on the flexible synthetic route using bromoallenes as described in Chap. 2, the author decided to explore a shorter total synthesis of pachastrissamine. This would introduce the alkyl side chain at the beginning. The author expected that palladium(0)-catalyzed cyclization of type **3** internal bromoallenes or type **4** propargylic substrates, bearing hydroxy and benzamide groups as nucleophilic functional groups, could regio- and stereoselectively provide the desired bicyclic tetrahydrofuran **2** (Scheme 3.1). Further hydrogenation of the olefin in **2** from the convex face would allow creation of the C-2 chiral center. Initial examination has revealed that chemoselective preparation of type **3** 1,3-disubstituted bromoallenes and type **4** propargyl tosylates/bromides is difficult.[1] Therefore, the author chose type **4** propargyl carbonates and chlorides as potential substrates for the palladium(0)-catalyzed bis-cyclization reaction.

Initially, the author planned to synthesize the diastereomeric propargyl carbonates *syn*- and *anti*-**8** to investigate the difference in reactivity between the diastereoisomers (Scheme 3.2). Alkynol *syn*-**6** was prepared from (*S*)-Garner's aldehyde **5** following the literature [1]. The alkynol *syn*-**6** was converted into the corresponding carbonate *syn*-**7** by treatment with $ClCO_2Me$, pyridine and DMAP. Removal of the Boc and acetal groups with TFA and MeOH followed by acylation

[1] For example, treatment of propargyl mesylates with $CuBr{\cdot}SMe_2$ in the presence of LiBr gave a mixture of allenyl/propargyl bromides in low yield. Furthermore, deprotection of propargyl tosylates/bromides with TFA and MeOH followed by acylation with BzCl and $(i\text{-}Pr)_2NEt$ did not afford the desired benzamide.

S. Inuki, *Total Synthesis of Bioactive Natural Products by Palladium-Catalyzed Domino Cyclization of Allenes and Related Compounds*, Springer Theses,
DOI: 10.1007/978-4-431-54043-4_3, © Springer 2012

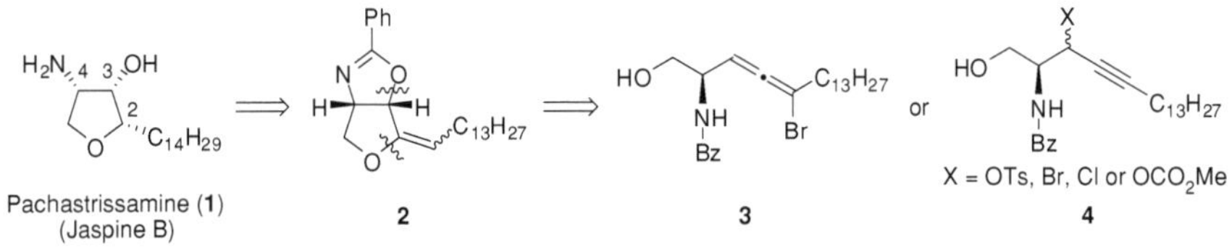

H_2N 4 3 OH 2 $C_{14}H_{29}$ O

Pachastrissamine (**1**)
(Jaspine B)

Ph N O H H $C_{13}H_{27}$ O

2

HO $C_{13}H_{27}$ NH Bz Br

3

or

X HO NH Bz $C_{13}H_{27}$

X = OTs, Br, Cl or OCO_2Me

4

Scheme 3.1 Retrosynthetic analysis of pachastrissamine (**1**)

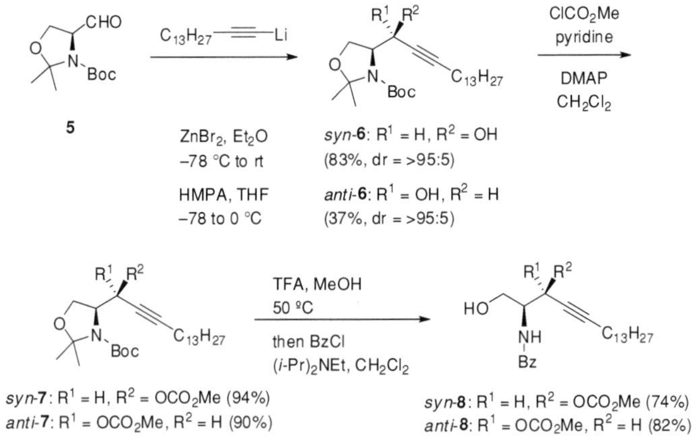

CHO O N Boc

5

$C_{13}H_{27}$—≡—Li

$ZnBr_2$, Et_2O
−78 °C to rt

HMPA, THF
−78 to 0 °C

R^1 R^2 O N Boc $C_{13}H_{27}$

syn-**6**: R^1 = H, R^2 = OH
(83%, dr = >95:5)

anti-**6**: R^1 = OH, R^2 = H
(37%, dr = >95:5)

$ClCO_2Me$
pyridine
DMAP
CH_2Cl_2

R^1 R^2 O N Boc $C_{13}H_{27}$

syn-**7**: R^1 = H, R^2 = OCO_2Me (94%)
anti-**7**: R^1 = OCO_2Me, R^2 = H (90%)

TFA, MeOH
50 °C
then BzCl
(i-Pr)$_2$NEt, CH_2Cl_2

R^1 R^2 HO NH Bz $C_{13}H_{27}$

syn-**8**: R^1 = H, R^2 = OCO_2Me (74%)
anti-**8**: R^1 = OCO_2Me, R^2 = H (82%)

Scheme 3.2 Synthesis of propargyl carbonates *syn*- and *anti*-**8**

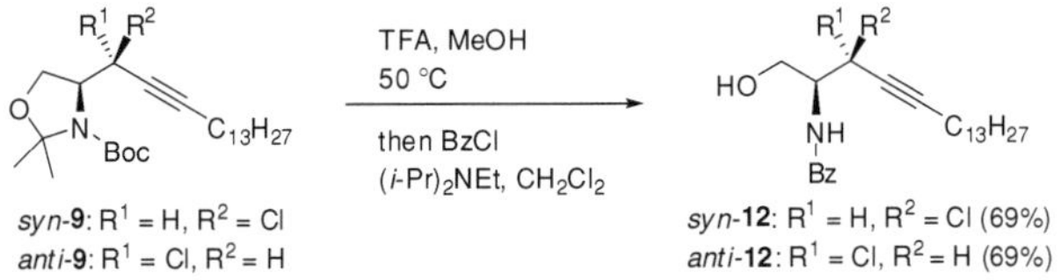

R^1 R^2 O N Boc $C_{13}H_{27}$

syn-**9**: R^1 = H, R^2 = Cl
anti-**9**: R^1 = Cl, R^2 = H

TFA, MeOH
50 °C
then BzCl
(i-Pr)$_2$NEt, CH_2Cl_2

R^1 R^2 HO NH Bz $C_{13}H_{27}$

syn-**12**: R^1 = H, R^2 = Cl (69%)
anti-**12**: R^1 = Cl, R^2 = H (69%)

Scheme 3.3 Synthesis of propargyl chlorides *syn*- and *anti*-**12**

with BzCl and (i-Pr)$_2$NEt gave the benzamide *syn*-**8**. The isomeric benzamide *anti*-**8** was obtained in the same manner via the alkynol *anti*-**6**[2] derived from (*S*)-Garner's aldehyde **5.**

The author next examined preparation of the required propargyl chloride by chlorination of propargyl alcohol **6** (Table 3.1) (fluorination reaction of a similar

[2] According to the literature [1], *anti*-**6** was produced in 71% yield. However, in this study, the desired product was obtained in low yield (37%) along with unidentified side products, and the optical rotation of alkynol *anti*-**6** was slightly decreased: $[\alpha]_D^{25}$ −33.6 (c 1.33, $CHCl_3$) [lit $[\alpha]_D^{25}$ −40.1 (c 1.0, $CHCl_3$)].

Table 3.1 Chlorination of propargyl alcohols [a]

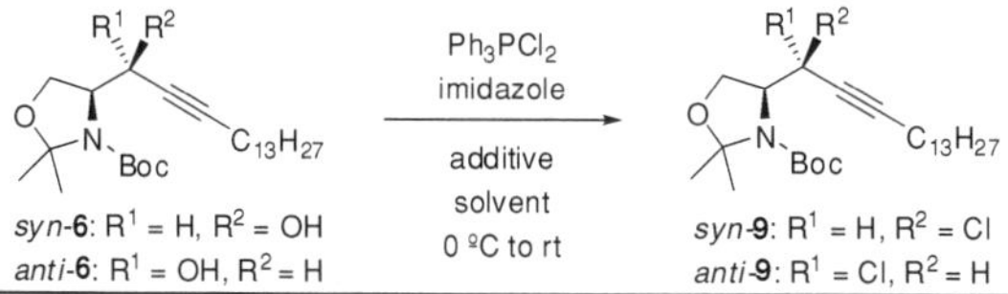

syn-**6**: R^1 = H, R^2 = OH
anti-**6**: R^1 = OH, R^2 = H

Ph_3PCl_2, imidazole, additive, solvent, 0 °C to rt

syn-**9**: R^1 = H, R^2 = Cl
anti-**9**: R^1 = Cl, R^2 = H

Entry	Substr.	Additive	Solvent	Yield (%)[b]	dr (*syn*:*anti*)[c]
1	*syn*-**6**	–	DMF	9	>95:5
2	*syn*-**6**	–	MeCN	15	93:7
3	*syn*-**6**	–	THF	22	>95:5
4	*syn*-**6**	–	CH_2Cl_2	30	>95:5
5	*syn*-**6**	–	Toluene	48	55:45
6	*syn*-**6**	LiCl[d]	CH_2Cl_2	8	>95:5
7	*syn*-**6**	$(n\text{-}Bu)_4NCl$[d]	CH_2Cl_2	14	>95:5
8	*anti*-**6**	–	CH_2Cl_2	47	5: > 95

[a] Reactions were carried out with Ph_3PCl_2 (4.0 equiv) and imidazole (4.0 equiv) for 2–8 h
[b] Isolated yields
[c] Determined by ^{1}H NMR analysis
[d] 4.0 equiv

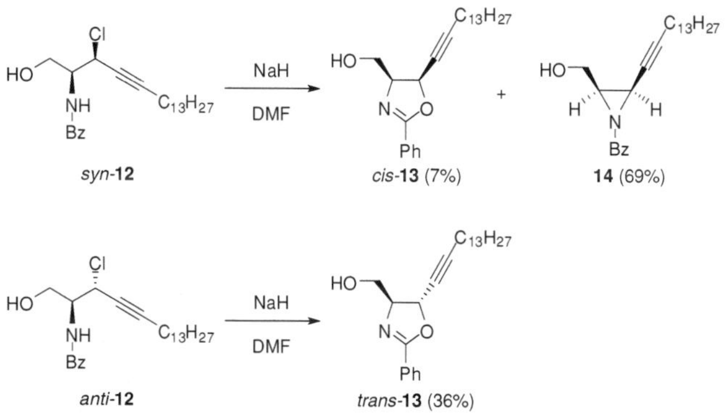

syn-**12** → NaH, DMF → *cis*-**13** (7%) + **14** (69%)

anti-**12** → NaH, DMF → *trans*-**13** (36%)

Scheme 3.4 Determination of the relative configuration of *syn*- and *anti*-**12**

alkynol is known to proceed with inversion of configuration, see Ref. [2]). Contrary to the author's expectations, the reaction of *syn*-**6** with Ph_3PCl_2 and imidazole in DMF afforded propargyl chloride *syn*-**9** in only 9% yield (entry 1). The *syn*-configuration of chloride, determined by cyclization of the corresponding benzamide **12** (vide infra, Schemes 3.3 and 3.4), demonstrates that the reaction proceeds with net retention of configuration (for related examples of Mitsunobu-type reaction with net retention of configuration by participation of a vicinal nitorogen functionality,

Table 3.2 Palladium-catalyzed cascade cyclization of propargyl chlorides [a]

syn-**12**: R^1 = H, R^2 = Cl
anti-**12**: R^1 = Cl, R^2 = H

Pd(PPh3)4, base, solvent, 50 °C → (*E*)-**2**, (*Z*)-**2**, *cis*-**13**, *trans*-**13**

Entry	Substr.	Base (equiv)	Solvent	2		13	
				Yield (%)[b]	*E*:*Z*[c]	Yield (%)[b]	*cis*:*trans*[c]
1	*syn*-**12**	NaH (2.5)	MeOH	49	92:8	ca. 12	80:20
2	*syn*-**12**	NaH (2.5)	THF	ca. 12	–	18	>95:5
3	*syn*-**12**	NaH (2.5)	THF:MeOH (10:1)	21	54:46	10	69:31
4[d]	*syn*-**12**	K_2CO_3 (1.2)	THF:MeOH (10:1)	24	>95:5	<18	77:22
5	*syn*-**12**	Cs_2CO_3 (1.2)	THF:MeOH (10:1)	73	95:5	<18	74:26
6[e]	*syn*-**12**	Cs_2CO_3 (1.2)	THF:MeOH (10:1)	89	>95:5	trace	–
7[e]	*anti*-**12**	Cs_2CO_3 (1.2)	THF:MeOH (10:1)	55	13:87	32	55:45

[a] Reactions were carried out with $Pd(PPh_3)_4$ (5 mol %) at 0.1 M for 1–1.5 h
[b] Isolated yields
[c] Determined by ^{1}H NMR analysis
[d] 46% of *syn*-**12** was recovered
[e] Reactions were carried out using 10 mol % of $Pd(PPh_3)_4$

see Refs. [3–6]).[3] Changing the solvent from DMF to MeCN, THF or CH_2Cl_2 increased the yields of the desired products to some extent with high diastereoselectivities (entries 2–4). It should be noted that the use of toluene as solvent provided the desired propargyl chloride in moderate yield (48%), but with extremely low diastereoselectivity (55:45, entry 5). Further screening of the reaction conditions using the additives LiCl or $(n\text{-Bu})_4NCl$ did not enhance the yield of the desired product. When the alkynol *anti*-**6** was employed, propargyl chloride *anti*-**9** was similarly produced by net retention of configuration in 47% yield.

Next the author prepared benzamides *syn*- and *anti*-**12** by removal of the Boc and acetal groups with TFA and MeOH, followed by acylation with BzCl and $(i\text{-Pr})_2NEt$ (Scheme 3.3). The relative configuration of *syn*- and *anti*-**12** was

[3] The observed stereoretention in the chlorination can be rationalized by the double inversion pathway. Activation of *syn*-**6** by Ph_3PCl_2 followed by initial intramolecular nucleophilic attack of Boc group to the activated propargylic position would form bicyclic intermediate **11** through inversion of configuration. The second intermolecular nucleophilic attack of chloride anion via stereoinversion then gives *syn*-**9**.

R = $C_{13}H_{27}$; *syn*-**6** → (Ph_3PCl_2) → **10** → (inversion) → **11** → (Cl^-, inversion) → *syn*-**9**

determined by derivatization to the corresponding oxazolines or aziridines (Scheme 3.4). The chloride *syn*-**12** was subjected to NaH in DMF to give the oxazoline *cis*-**13**[4] (7%) and aziridine *cis*-**14**[5] (69%). In contrast, the reaction of *anti*-**12** gave the oxazoline *trans*-**13**[6] in 36% yield.

The author investigated cascade cyclization of propargyl chlorides *syn*-**12** and *anti*-**12** in the presence of palladium(0) (Table 3.2). Reaction of *syn*-**12** with $Pd(PPh_3)_4$ (5 mol %) and NaH (2.5 equiv) in MeOH at 50°C (standard conditions for cyclization of propargyl bromide) [7, 8] (for a related work, see Ref. [9]) afforded the desired bicyclic tetrahydrofuran **2** in 49% yield with high *E*-selectivity (E:Z[7] = 92:8, entry 1). Although the undesired mono-cyclized furan derivatives were not obtained [7, 8] (for a related work, see Ref. [9]), S_N2-type oxazoline

---

[4] The relative configuration of *cis*-**13** was confirmed by comparison with the authentic sample prepared from the known alkynol *syn*-**6** [1].

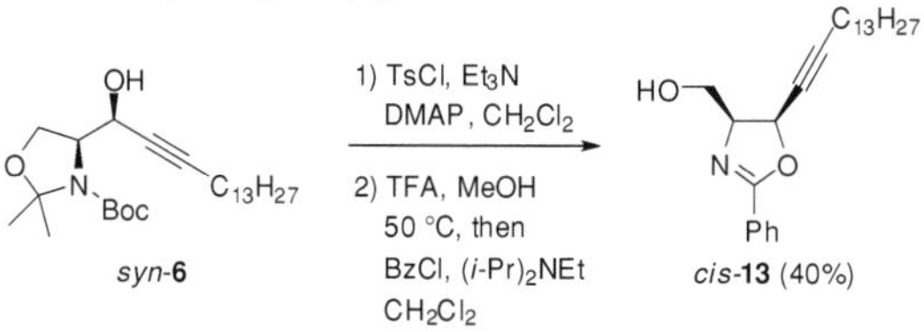

OH
N Boc
$C_{13}H_{27}$
syn-**6**
1) TsCl, Et_3N DMAP, CH_2Cl_2
2) TFA, MeOH 50 °C, then BzCl, $(i\text{-Pr})_2NEt$ CH_2Cl_2
HO
$C_{13}H_{27}$
N O
Ph
cis-**13** (40%)

[5] The relative configuration of *cis*-**14** was determined using a J_{Hab}-based configurational analysis: the observed H_a–H_b coupling constant (J_{Hab} = 6.0 Hz) indicates the 2,3-*cis* configuration of the aziridine [10]

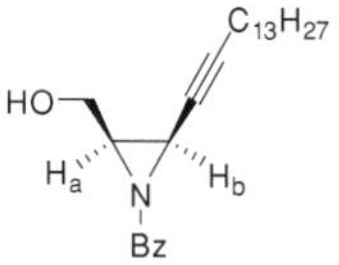

$C_{13}H_{27}$
HO
H_a N H_b
Bz
14
J_{Hab} = 6.0 Hz

[6] The relative configuration of *trans*-**13** was confirmed by comparison with the authentic sample prepared from the known alkynol *anti*-**6** [1].

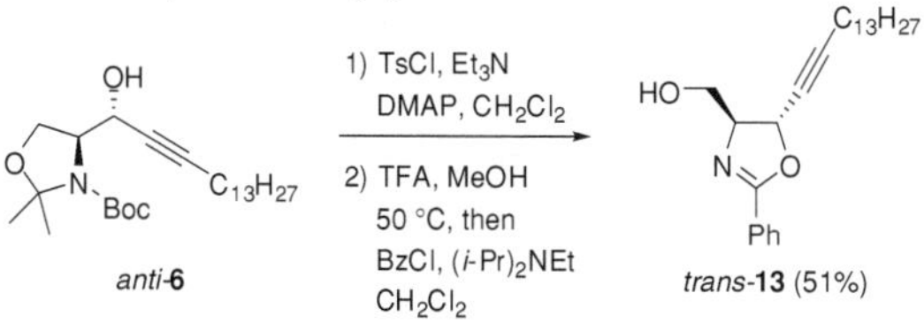

OH
N Boc
$C_{13}H_{27}$
anti-**6**
1) TsCl, Et_3N DMAP, CH_2Cl_2
2) TFA, MeOH 50 °C, then BzCl, $(i\text{-Pr})_2NEt$ CH_2Cl_2
HO
$C_{13}H_{27}$
N O
Ph
trans-**13** (51%)

[7] The configuration of the bicyclic tetrahydrofuran **2** was determined by NOE analysis.

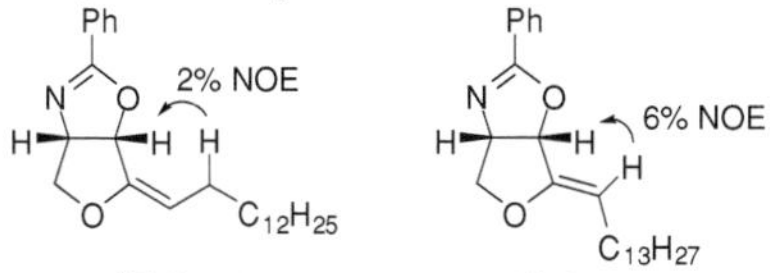

Ph
N O
H H H
2% NOE
O
$C_{12}H_{25}$
(*E*)-**2**

Ph
N O
H H
6% NOE
H
O
$C_{13}H_{27}$
(*Z*)-**2**

Table 3.3 Palladium-catalyzed cascade cyclization of propargyl carbonates[a]

$Pd(PPh_3)_4$, base, additive, solvent, 50 °C

syn-**8**: R^1 = H, R^2 = OCO_2Me
anti-**8**: R^1 = OCO_2Me, R^2 = H

(*E*)-**2** (*Z*)-**2** *cis*-**13** *trans*-**13**

Entry	Substr.	Base[b]	Solvent	**2** Yield (%)[c,d]	*E*:*Z*[e]	**13** Yield (%)[c,d]	*cis*:*trans*[e]
1	*syn*-**8**	–	MeOH	2[f]	>95:5	ND	–
2	*syn*-**8**	–	THF	69	>95:5	ND	–
3[g]	*syn*-**8**	–	THF	60	>95:5	ND	–
4	*syn*-**8**	–	THF:MeOH (10:1)	65	>95:5	ND	–
5	*syn*-**8**	Cs_2CO_3	THF	67–78	>95:5	ND	–
6[g]	*syn*-**8**	Cs_2CO_3	THF	14	>95:5	ND	–
7	*anti*-**8**	–	THF	<20	>95:5	60	82:18
8[h]	*anti*-**8**	–	THF	ND	–	ND	–
9[i]	*anti*-**8**	–	THF	ND	–	ND	–
10	*anti*-**8**	–	THF:MeOH (10:1)	39	13:87	15	>95:5
11	*anti*-**8**	Cs_2CO_3	THF	<26	42:58	39	>95:5
12	*anti*-**8**	Cs_2CO_3	THF:MeOH (10:1)	7	16:84	ND	–

[a] Reactions were carried out with $Pd(PPh_3)_4$ (5 mol %) at 0.1 M for 2–4.5 h
[b] 1.2 equiv of a base were used
[c] Isolated yields
[d] ND = Not detected
[e] Determined by ^1H NMR analysis
[f] Solvolysis product was obtained
[g] Reactions were carried out on 1 g scale
[h] $(n\text{-}Bu)_4NCl$ was used as an additive (0.3 equiv)
[i] LiCl was used as an additive (1.0 equiv)

product **13** was observed (ca. 12%, *cis*:*trans* = 80:20).[8] Changing the solvent from MeOH to THF or THF/MeOH (10:1) did not enhance the yield of the desired product (entries 2, 3). Of the several bases investigated, Cs_2CO_3 was the most effective yielding 73% of **2** as a 95:5 *E*/*Z* mixture (entries 3–5). Furthermore, increased loading of $Pd(PPh_3)_4$ (10 mol %) improved the yield to 89% and suppressed formation of oxazoline **13** (entry 6). When *anti*-**12** was subjected to the optimized reaction conditions (entry 6) the desired bicyclic tetrahydrofuran **2** was obtained in 55% yield with moderate *Z*-selectivity (*E*:*Z* = 13:87), along with oxazoline **13** (32%, *cis*:*trans* = 55:45, entry 7). These results show utility of

[8] The minor isomer *trans*-**13** could be produced by double inversion pathway: *anti*-attack of benzamide group to propargyl/allenyl palladium complex, formed by *anti*-attack of palladium(0) to *syn*-**12**, will produce the net retention product *trans*-**13**.

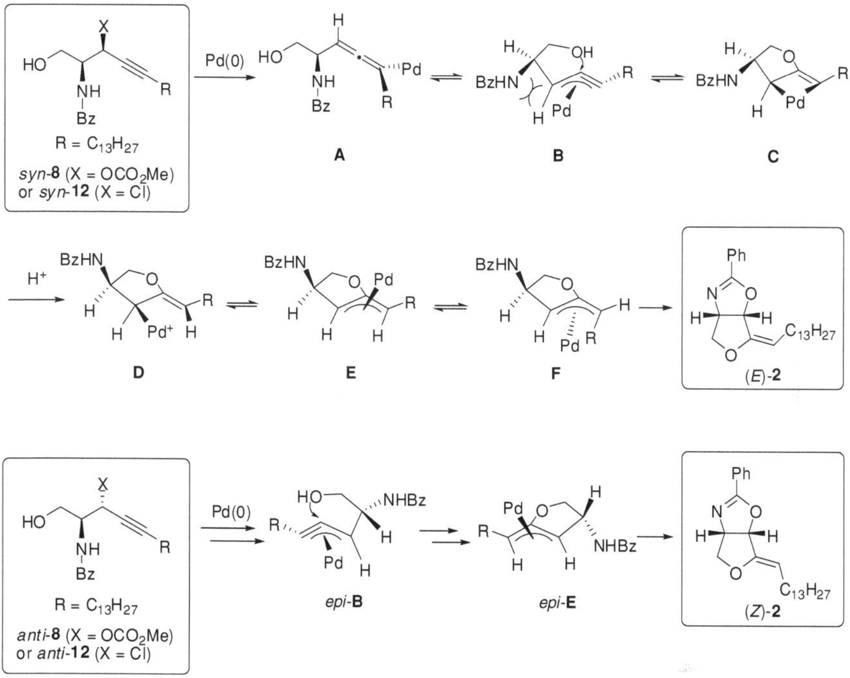

Scheme 3.5 Proposed mechanism for cascade reaction

propargyl chlorides with *syn*-configuration as a precursor of bicyclic products, and a clear difference in reactivity between the diastereomeric substrates.

The author next investigated the reaction of propargyl carbonates *syn*-**8** and *anti*-**8** in the presence of palladium(0) (Table 3.3). Treatment of *syn*-**8** with $Pd(PPh_3)_4$ (5 mol %) in MeOH at 50°C afforded the desired bicyclic tetrahydrofuran **2** in low yield (2%) (entry 1). The main product was the corresponding diol formed by alcoholysis of carbonate **8** (entry 1). When THF was used as the reaction solvent, **2** was obtained with a higher yield (69%) and excellent *E*-selectivity (*E*:*Z* = >95:5, entry 2). Conducing the reaction on a 1 g scale also gave the desired product in satisfactory yield (60%, entry 3). According to the previous reports [7–9, 11–14] solvents containing alcohol promote the palladium-catalyzed reactions of bromoallenes or propargylic compounds. However, the addition of MeOH did not improve the yield (entry 4). Although the reaction of *syn*-**8** on a 40 mg scale in the presence of Cs_2CO_3 gave **2** in 67–78% yield, this was not reproducible on the 1 g scale (entries 5, 6). Next the diastereomeric carbonates *anti*-**8** was reacted under the above optimized conditions (entry 2). This gave the desired product **2** in unexpectedly low yield (< 20%) with >95:5 *E*-selectivity, and S_N2 product **13** (60%, *cis*:*trans* = 82:18) (entry 7). This result is quite different from the reaction using propargyl chlorides (Table 3.2, entries 6, 7). To achieve efficient transformation, further screening was carried out based on the examination of propargyl chlorides (Table 3.2). Addition of a chloride anion source such as $(n\text{-Bu})_4NCl$ or LiCl did not

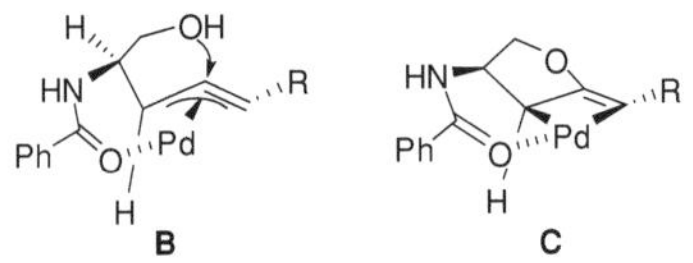

Fig. 3.1 Possible effects of the amide group in the intermediates **B** and **C**

afford **2** (entries 8, 9). Among the several reaction conditions investigated (entries 10–12), the use of a mixed solvent THF/MeOH (10:1) under base-free conditions gave the most efficient conversion of *anti*-**8** into the desired product **2** in favor of Z-isomer (39%, *E*:*Z* = 13:87, entry 10). These results indicate that an alcoholic solvent plays an important role in stereospecific cyclization of some propargylic systems.

Formation of (*E*)-**2** from the carbonate *syn*-**8** or chloride *syn*-**12** can be explained as follows (Scheme 3.5). Initially, regio- and stereo-selective S_N2' attack of palladium(0) to propargylic compounds proceeds to yield the allenylpalladium intermediate **A**. First cyclization by the hydroxy group on the central carbon of η^3-propargylpalladium complex **B** [15–17], which is formed by rearrangement of **A**, would generate a fused palladacyclobutene intermediate **C** [18, 19] (formation of a palladacyclobutene intermediate in a related reaction has been well rationalized by DFT calculation, see Ref. [20]). This is followed by protonation to form complex **D** without loss of chirality. After formation of the η^3-allylpalladium intermediate **E**, isomerization to *anti*-type complex **F** is necessary for the next *anti*-cyclization. Therefore, transformation into the intermediate **F** through η^3-η^1-η^3 equilibration followed by the second cyclization by the benzamide group then gives (*E*)-**2** (for related chirality transfer in the reaction of the propargylic compounds via palladacyclobutene intermediates, see Ref. [21]). The carbonate *anti*-**8** or the chloride *anti*-**12** would be converted into η^3-propargylpalladium complex *epi*-**B** via S_N2' attack of palladium(0). Cyclization by hydroxy group and subsequent protonation will form η^3-allylpalladium intermediate *epi*-**E**, which gives (*Z*)-**2** by *anti* attack of the benzamide group. It should be noted that the reaction of *syn*-propargylic compounds, which would have unfavorable steric interaction between the palladium and the benzamide group in the first cyclization step, proceeds more efficiently than that of *anti*-compounds (entries 6 vs. 7, Table 3.2; entries 2 vs. 7, Table 3.3). This result suggests that coordination of the benzamide group to palladium would promote the first cyclization by stabilizing the reactive conformer as depicted in **B** and/or the resulting palladacyclobutene intermediate **C** (Fig. 3.1). Although the exact reason for the lower *Z*-selectivities in reaction of the *anti*-substrates (Table 3.3, entries 7, 10–12) is unclear, it can be attributed to epimerization of allenylpalladium or η^3-propargylpalladium complex *epi*-**B** due to slower cyclization without assistance of a coordinating effect [19, 22–28].

The author next investigated hydrogenation of (*E*)-olefin **2**, which enabled creation of the C-2 stereogenic center (Table 3.4). When using 10% Pd/C, the desired product **15** was obtained in 45% yield as the sole diastereomer.

Table 3.4 Hydrogenation of (*E*)-olefin **2** [a]

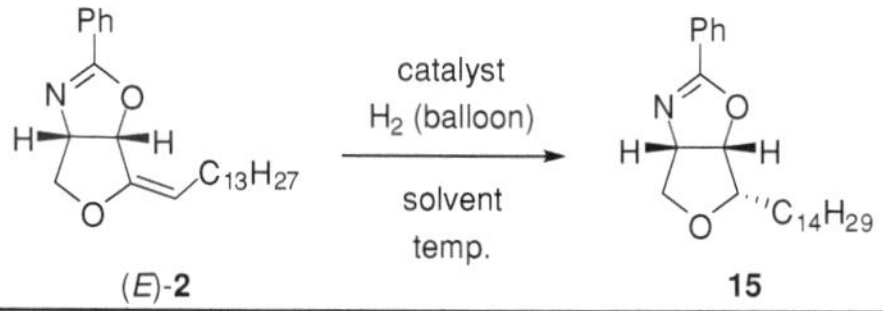

Entry	Catalyst (mol %)	Solvent	Temp. (°C)	Time (h)	Yield (%)[b]
1	10% Pd/C (5)	EtOAc	rt	1	45
2	$Pd(OH)_2/C$ (5)	EtOAc	rt	16	5
3	PtO_2 (5)	EtOAc	rt	22	ca. 5
4	Ir-black (10)	EtOAc	50	25	28
5	$(Ph_3P)_3RhCl$ (5)	C_6H_6/EtOH	50	27	62
6	$(Ph_3P)_3RhCl$ (10)	C_6H_6/EtOH	50	25	82
7	Crabtree cat. (10)	DCE	70	21	30

[a] Reactions were carried out with pure (*E*)-olefin **2**
[b] Isolated yields. DCE = 1,2-dichloroethane

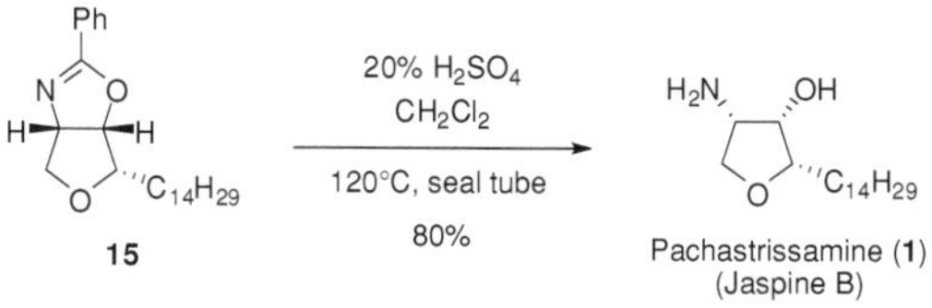

Scheme 3.6 Hydrolysis of oxazoline ring

Heterogeneous catalysts ($Pd(OH)_2/C$, PtO_2 and Ir-black, entries 2–4) were screened further but the yield of the desired product decreased with a prolonged reaction time. On examination of the homogenous catalyst, the author found 5 mol % of $(Ph_3P)_3RhCl$ enhanced the yield (62%). When the catalyst loading was increased to 10 mol %, the desired product **15** was isolated in a higher yield (82%, entry 6). In contrast, use of Crabtree catalyst [29, 30] decreased the yield of **15**–30% (entry 7).

After synthesis of the pachastrissamine derivatives **15** bearing the requisite functionalities, the author tested cleavage of the oxazoline ring. The hydrolysis of **15** under harsh conditions (20% H_2SO_4, CH_2Cl_2, sealed tube, 120°C) gave the desired conversion in 80% yield (Scheme 3.6) [31]. Next, the author decided to develop an alternative approach for oxazoline group cleavage, and used a two-step reduction under mild conditions [32–36]. Treatment of **15** with DIBAL-H successfully produced the desired benzyl protected pachastrissamine **16** quantitatively [34]. Finally, removal of the benzyl group with $Pd(OH)_2/C$ led to pachastrissamine **1** in 86% yield (Scheme 3.7).

In conclusion, the author has developed a novel ring-construction/stereoselective functionalization cascade by palladium(0)-catalyzed bis-cyclization of propargylic carbonates and chlorides. When using propargylic compounds, a reactivity difference was observed between the diastereomeric *syn*- or *anti*-substrates. Reaction of the *syn*-propargylic isomer proceeded more efficiently than the

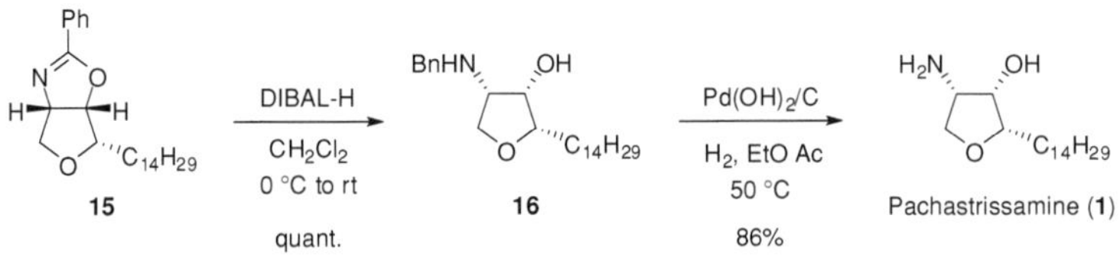

Scheme 3.7 Cleavage of oxazoline ring

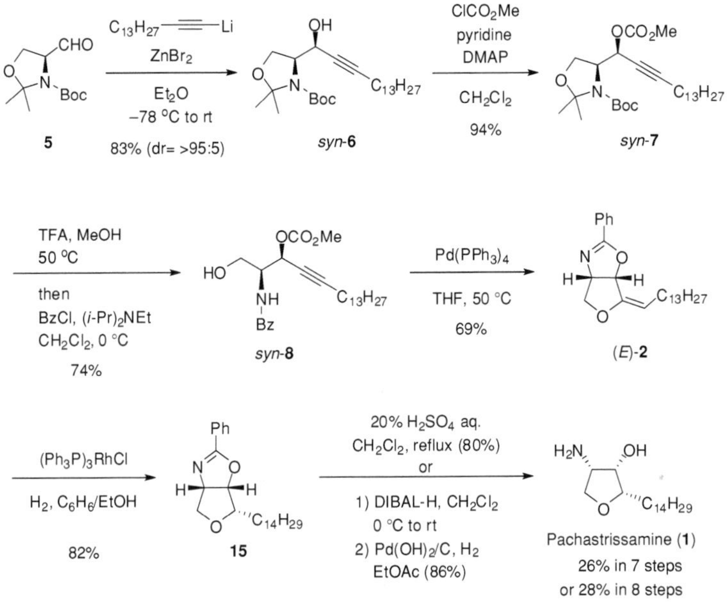

Scheme 3.8 Straightforward total synthesis of pachastrissamine (1)

corresponding *anti*-isomer. The author has achieved a short-step total synthesis of pachastrissamine using propargylic carbonates. This synthetic route furnishes pachastrissamine in 26% overall yield in seven steps (final deprotection by hydrolysis) or 28% overall yield in eight steps (reductive deprotection) starting from Garner's aldehyde as the sole chiral source (Scheme 3.8).

3.1 Experimental Section

3.1.1 General Methods

All moisture-sensitive reaction were performed using syringe-septum cap techniques under an argon atmosphere and all glassware was dried in an oven at 80°C for 2 h prior to use. Reactions at −78°C employed a CO_2–MeOH bath. Melting points were measured by a hot stage melting point apparatus (uncorrected). Optical rotations were measured with a JASCO P-1020 polarimeter. For flash chromatography, Wakosil C-300, Wakogel C-300E or Chromatorex® was employed. ^{1}H NMR spectra were recorded using a JEOR AL-400 or JEOL ECA-500 spectrometer, and chemical shifts are reported in δ (ppm) relative to TMS (in $CDCl_3$) as internal standard. ^{13}C NMR spectra were recorded using a JEOL AL-400 or JEOL ECA-500 spectrometer and referenced to the residual $CHCl_3$ signal. ^{1}H NMR spectra are tabulated as follows: chemical shift, multiplicity (b = broad, s = singlet, d = doublet, t = triplet, q = quartet, m = multiplet), number of protons, and coupling constant(s). Exact mass (HRMS) spectra were recorded on a JMS-HX/HX 110A mass spectrometer. Infrared (IR) spectra were obtained on a JASCO FT/IR-4100 FT-IR spectrometer with JASCO ATR PRO410-S.

3.1.2 tert-Butyl (S)-4-[(S)-1-Hydroxyhexadec-2-yn-1-yl]-2, 2-dimethyloxazolidine-3-carboxylate (syn-6)

To a solution of pentadec-1-yne (562 mg, 2.70 mmol) in Et_2O (5 mL) was added dropwise *n*-BuLi in hexane (1.6 M; 1.64 mL, 2.61 mmol) at −20°C. After the resulting white suspension was stirred for 1 h at this temperature, a solution of $ZnBr_2$ in Et_2O (ca. 1.0 M; 2.78 mL, 2.78 mmol) was added at 0°C. After stirring for 1 h at this temperature and for 1 h at room temperature, a solution of Garner's aldehyde **5** (200 mg, 0.87 mmol) in Et_2O (0.75 mL) was added dropwise at −78°C. The mixture was allowed to warm to room temperature. After stirring for 12 h at this temperature, the mixture was quenched by addition of saturated NH_4Cl at −20°C. After dilution with H_2O, aqueous layer was separated and extracted with Et_2O. The combined Et_2O extracts were washed with brine, and dried over $MgSO_4$. The filtrate was concentrated under reduced pressure to give a colorless oil, which was purified by flash chromatography over silica gel with *n*-hexane–EtOAc (7:1) to give *syn*-**6** as a colorless oil (315 mg, 83% yield): $[\alpha]_D^{25} - 32.3$ (*c* 1.29, $CHCl_3$) [lit $[\alpha]_D^{25} - 32.4$ (*c* 1.3, $CHCl_3$)]. All the spectral data were in agreement with those reported by Herold [1].

3.1.3 tert-Butyl (S)-4-[(R)-1-Hydroxyhexadec-2-yn-1-yl]-2, 2-dimethyloxazolidine-3-carboxylate (anti-6)

To a solution of pentadec-1-yne (9.55 g, 45.8 mmol) in THF (125 mL) was added dropwise *n*-BuLi in hexane (1.6 M; 27.4 mL, 43.6 mmol) at −20°C. After the resulting white suspension was stirred for 2.0 h at this temperature, HMPA (11.0 mL, 63.2 mmol) was added at this temperature. After stirring for 10 min at this temperature, a solution of Garner's aldehyde **5** (5.00 g, 21.8 mmol) in THF

(18.0 mL) was added dropwise at −78°C. After stirring for 10 min at this temperature, the mixture was allowed to warm to −20°C. The mixture was stirred for 3.0 h at −20°C and quenched with saturated NH_4Cl. The whole was extracted with Et_2O and the extract was washed with H_2O and brine, and was dried over Na_2SO_4. Concentration under reduced pressure followed by flash chromatography over silica gel with *n*-hexane–EtOAc (5:1) gave *anti*-**6** as a pale yellow oil (3.51 g, 37% yield): $[\alpha]_D^{25}$ −33.6 (*c* 1.33, $CHCl_3$) [lit $[\alpha]_D^{25}$ −40.1 (*c* 1.0, $CHCl_3$)]. All the spectral data except for optical rotation were in agreement with those reported by Herold [1].

3.1.4 *tert-Butyl (S)-4-[(S)-1-(Methoxycarbonyloxy) hexadec-2-yn-1-yl]-2,2-dimethyloxazolidine -3-carboxylate (syn-7)*

To a stirred solution of *syn*-**6** (1.00 g, 2.28 mmol) in CH_2Cl_2 (8.0 mL) were added pyridine (1.11 mL, 13.7 mmol), DMAP (55.7 mg, 0.46 mmol) and $ClCO_2Me$ (1.06 mL, 13.7 mmol) at 0°C, and the mixture was stirred for 1.5 h at room temperature, followed by quenching with saturated NH_4Cl. The whole was extracted with EtOAc. The extract was washed with H_2O and brine, dried over $MgSO_4$, and concentrated under pressure to give an oily residue, which was purified by column chromatography over silica gel with *n*-hexane–EtOAc (11:1) to give *syn*-**7** as a colorless oil (1.06 g, 94% yield): $[\alpha]_D^{25}$ − 23.4 (*c* 1.10, $CHCl_3$); IR (neat): 2247 (C≡C), 1757 (C=O), 1705 (C=O); ^{1}H NMR (500 MHz, DMSO, 100°C) δ 0.86 (t, J = 6.9 Hz, 3H), 1.22–1.38 (m, 20H), 1.43 (s, 3H), 1.43 (s, 9H), 1.44–1.47 (m, 2H), 1.52 (s, 3H), 2.20 (td, J = 6.9, 2.3 Hz, 2H), 3.72 (s, 3H), 3.98 (dd, J = 9.2, 3.4 Hz, 1H), 3.99–4.02 (m, 1H), 4.02–4.07 (m, 1H), 5.59 (dt, J = 5.1, 2.3 Hz, 1H); ^{13}C NMR (125 MHz, $CDCl_3$; as a mixture of amide rotamers) δ 14.1, 18.8, 22.6, 23.4, 24.7, 26.0, 26.8, 28.2, 28.3, 28.8, 29.0, 29.3 (2C), 29.4, 29.6 (3C), 31.8, 54.9, 58.7 (0.5C), 59.3 (0.5C), 64.4 (0.5C), 64.7 (0.5C), 68.0 (0.5C), 68.4 (0.5C), 73.9, 80.5, 88.6 (0.5C), 89.0 (0.5C), 94.3 (0.5C), 94.9 (0.5C), 151.7, 154.5 (0.5C), 154.6 (0.5C). *Anal.* Calcd for $C_{28}H_{49}NO_6$: C, 67.84; H, 9.96; N, 2.83. Found: C, 67.90; H, 9.68; N, 2.79.

3.1.5 *tert-Butyl (S)-4-[(R)-1-(Methoxycarbonyloxy) hexadec-2-yn-1-yl]-2,2-dimethyloxazolidine -3-carboxylate (anti-7)*

By a procedure identical with that described for synthesis of *syn*-**7** from *syn*-**6**, the propargyl alcohol *anti*-**6** (500 mg, 1.14 mmol) was converted into *anti*-**7** as a colorless oil (507 mg, 90% yield): $[\alpha]_D^{26}$ −62.9 (*c* 1.49, $CHCl_3$); IR (neat): 2236

(C≡C), 1755 (C=O), 1707 (C=O); ^{1}H NMR (500 MHz, DMSO, 100°C) δ 0.86 (t, J = 6.6 Hz, 3H), 1.22–1.38 (m, 20H), 1.39–1.43 (m, 15H), 1.44–1.49 (m, 2H), 2.21 (td, J = 6.9, 1.7 Hz, 2H), 3.71 (s, 3H), 4.02–4.05 (m, 3H), 5.61–5.64 (m, 1H); ^{13}C NMR (125 MHz, $CDCl_3$; as a mixture of amide rotamers) δ 14.1, 18.7, 22.7, 23.3, 24.6, 25.4, 26.2, 28.2, 28.3, 28.8, 29.0, 29.3, 29.5, 29.6 (3C), 29.7, 31.9, 54.8 (0.5C), 54.9 (0.5C), 60.1 (0.5C), 60.4 (0.5C), 63.7 (0.5C), 64.2 (0.5C), 66.9 (0.5C), 67.3 (0.5C), 74.6, 80.5 (0.5C), 80.6 (0.5C), 88.8 (0.5C), 89.2 (0.5C), 94.4 (0.5C), 95.1 (0.5C), 151.5 (0.5C), 152.3 (0.5C), 155.1. *Anal.* Calcd for $C_{28}H_{49}NO_6$: C, 67.84; H, 9.96; N, 2.83. Found: C, 67.55; H, 9.79; N, 2.81.

3.1.6 (2S,3S)-2-Benzamido-1-hydroxyoctadec-4-yn-3-yl Methyl Carbonate (syn-8)

To a stirred solution of *syn*-**7** (963 mg, 1.94 mmol) in MeOH (6.5 mL) at 0°C was added trifluoroacetic acid (18 mL), and the mixture was stirred for 1.5 h at 50°C. The mixture was concentrated under reduced pressure, and the residue was dissolved in CH_2Cl_2 (18 mL). The solution was made neutral with $(i\text{-Pr})_2NEt$ at 0°C. $(i\text{-Pr})_2NEt$ (1.18 mL, 6.79 mmol) and BzCl (0.248 mL, 2.13 mmol) were added to the mixture under stirring at 0°C. The mixture was stirred for 2.5 h at this temperature, followed by addition of H_2O. The whole was extracted with EtOAc. The extract was washed successively with 1 N HCl, H_2O and brine, and dried over Na_2SO_4. The filtrate was concentrated under reduced pressure to give an oily residue, which was purified by flash chromatography over silica gel with *n*-hexane–EtOAc (3:2) to give *syn*-**8** as a pale yellow oil (664 mg, 74% yield): $[\alpha]_D^{25}$ + 33.3 (*c* 1.43, $CHCl_3$); IR (neat): 3380 (OH), 2237 (C≡C), 1755 (C=O), 1650 (C=O); ^{1}H NMR (500 MHz, $CDCl_3$) δ 0.88 (t, J = 6.6 Hz, 3H), 1.19–1.35 (m, 20H), 1.47 (tt, J = 7.2, 7.2 Hz, 2H), 2.20 (td, J = 7.2, 1.7 Hz, 2H), 2.55 (br s, 1H), 3.80 (dd, J = 11.5, 5.2 Hz, 1H), 3.81 (s, 3H), 4.02 (dd, J = 11.5, 4.6 Hz, 1H), 4.45–4.52 (m, 1H), 5.60–5.66 (m, 1H), 6.66 (d, J = 8.0 Hz, 1H), 7.44 (dd, J = 7.4, 7.4 Hz, 2H), 7.52 (t, J = 7.4 Hz, 1H), 7.80 (d, J = 7.4 Hz, 2H); ^{13}C NMR (125 MHz, $CDCl_3$) δ 14.1, 18.7, 22.7, 28.2, 28.8, 29.1, 29.3, 29.4, 29.6 (3C), 29.7, 31.9, 54.4, 55.2, 61.8, 67.5, 74.3, 89.7, 127.1, 128.6 (2C), 131.8 (2C), 133.9, 154.9, 167.8; HRMS (FAB) calcd for $C_{27}H_{42}NO_5$ $[M+H]^+$, 460.3063, found: 460.3068.

3.1.7 (2S,3R)-2-Benzamido-1-hydroxyoctadec-4-yn-3-yl Methyl Carbonate (anti-8)

By a procedure identical with that described for synthesis of *syn*-**8** from *syn*-**7**, the propargyl carbonate *anti*-**7** (438 mg, 0.88 mmol) was converted into *anti*-**8** as a colorless oil (331 mg, 82% yield): $[\alpha]_D^{26}$ −41.2 (*c* 1.15, $CHCl_3$); IR (neat): 3355 (OH), 2238 (C≡C), 1756 (C=O), 1651 (C=O); ^{1}H NMR (500 MHz, $CDCl_3$) δ 0.88 (t, J = 6.6 Hz, 3H), 1.23–1.30 (m, 18H), 1.32–1.39 (m, 2H), 1.51 (tt, J = 7.5, 7.5 Hz, 2H), 2.25 (td, J = 7.5, 1.3 Hz, 2H), 3.81 (s, 3H), 3.84 (dd, J = 12.0, 5.2 Hz, 1H), 4.07 (dd, J = 12.0, 4.3 Hz, 1H), 4.46–4.52 (m, 1H), 5.59–5.63 (m, 1H), 6.76 (d, J = 8.0 Hz, 1H), 7.45 (dd, J = 7.4, 7.2 Hz, 2H), 7.53 (t, J = 7.2 Hz, 1H), 7.79 (d, J = 7.4 Hz, 2H); ^{13}C NMR (125 MHz, $CDCl_3$) δ 14.1, 18.7, 22.7, 28.3, 28.9, 29.1, 29.3, 29.5, 29.6 (3C), 29.7, 31.9, 54.0, 55.2, 61.9, 68.8, 73.6, 90.4, 127.1, 128.6 (2C), 131.8 (2C), 133.9, 154.7, 167.8; HRMS (FAB) calcd for $C_{27}H_{42}NO_5$ $[M+H]^+$, 460.3063, found: 460.3060.

3.1.8 *tert-Butyl (S)-4-[(R)-1-Chlorohexadec-2-yn-1-yl]-2, 2-dimethyloxazolidine-3-carboxylate (syn-9)*

To a stirred solution of *syn*-**6** (2.00 g, 4.57 mmol) and imidazole (1.25 g, 18.3 mmol) in CH_2Cl_2 (8.0 mL) was added a solution of Ph_3PCl_2 (6.09 g, 18.3 mmol) in CH_2Cl_2 (12 mL) at 0°C. After stirring for 1.0 h at this temperature, concentration under reduced pressure gave an oily residue, which was purified by flash chromatography over silica gel with *n*-hexane–EtOAc (30:1) to give *syn*-**9** as a colorless oil (618 mg, 30% yield); $[\alpha]_D^{24}$ − 96.7 (*c* 1.17, $CHCl_3$); IR (neat): 2246 (C≡C), 1694 (C=O); ^{1}H NMR (500 MHz, DMSO, 100°C) δ 0.86 (t, J = 6.9 Hz, 3H), 1.20–1.38 (m, 20H), 1.43 (s, 3H), 1.44 (s, 9H), 1.45–1.48 (m, 2H), 1.53 (s, 3H), 2.23 (td, J = 6.9, 2.3 Hz, 2H), 4.05–4.12 (m, 3H), 5.02–5.09 (m, 1H); ^{13}C NMR (125 MHz, $CDCl_3$; as a mixture of amide rotamers) δ 14.1, 18.9, 22.7, 23.5, 24.9, 25.9, 26.5, 28.2, 28.3, 28.4, 28.8, 29.1, 29.3, 29.5, 29.6 (2C), 29.7, 31.9, 48.1 (0.5C), 49.0 (0.5C), 61.9 (0.5C), 62.2 (0.5C), 64.5 (0.5C), 64.9 (0.5C), 75.0 (0.5C), 75.4 (0.5C), 80.6 (0.5C), 80.8 (0.5C), 89.0 (0.5C), 89.5 (0.5C), 94.8 (0.5C), 95.5 (0.5C), 151.5 (0.5C), 152.5 (0.5C); HRMS (FAB) calcd for $C_{26}H_{47}ClNO_3$ $[M+H]^+$, 456.3244, found: 456.3248 (Table 3.1, Entry 4).

3.1.9 *tert-Butyl (S)-4-[(S)-1-Chlorohexadec-2-yn-1-yl]-2, 2-dimethyloxazolidine-3-carboxylate (anti-9)*

By a procedure identical with that described for synthesis of *syn*-**9** from *syn*-**6**, the propargyl alcohol *anti*-**6** (1.00 g, 2.28 mmol) was converted into *anti*-**9** as a colorless oil (488 mg, 47% yield): $[\alpha]_D^{26}$ −19.8 (*c* 1.27, $CHCl_3$); IR (neat): 2235 (C≡C), 1693 (C=O); ^{1}H NMR (500 MHz, DMSO, 120°C) δ 0.86 (t, J = 6.6 Hz, 3H), 1.23–1.38 (m, 20H), 1.43 (s, 3H), 1.44 (s, 9H), 1.45–1.49 (m, 2H), 1.52 (s, 3H), 2.23 (td, J = 6.8, 2.3 Hz, 2H), 4.02 (dd, J = 9.7, 2.9 Hz, 1H), 4.06 (dd, J = 9.7, 6.3 Hz, 1H), 4.16–4.20 (m, 1H), 5.05–5.09 (m, 1H); ^{13}C NMR

(125 MHz, $CDCl_3$; as a mixture of rotamers) δ 14.1, 18.8, 18.9, 22.7, 23.6, 24.9, 25.5, 26.2, 28.3, 28.4, 28.5, 28.8, 29.1, 29.4, 29.5, 29.6, 29.7, 31.9, 49.2 (0.5C), 49.8 (0.5C), 61.3 (0.5C), 61.8 (0.5C), 64.3 (0.5C), 64.9 (0.5C), 76.1, 80.5 (0.5C), 80.8 (0.5C), 88.9 (0.5C), 89.2 (0.5C), 94.8 (0.5C), 95.6 (0.5C), 151.6 (0.5C), 152.6 (0.5C); HRMS (FAB) calcd for $C_{26}H_{47}ClNO_3$ $[M+H]^+$, 456.3244, found: 456.3243 (Table 3.1, Entry 8).

3.1.10 N-[(2S,3R)-3-Chloro-1-hydroxyoctadec-4-yn-2-yl] benzamide (syn-12)

By a procedure identical with that described for the synthesis of the benzamide *syn*-**8** from *syn*-**7**, the chloride *syn*-**9** (49.1 mg, 0.108 mmol) was converted into *syn*-**12** (31.3 mg, 69% yield): white waxy solid; mp 60–61°C; $[\alpha]_D^{24}$ – 22.7 (*c* 1.20, $CHCl_3$); IR (neat): 3335 (OH), 2237 (C≡C), 1653 (C=O); ^1H NMR (500 MHz, $CDCl_3$) δ 0.88 (t, J = 6.9 Hz, 3H), 1.19–1.35 (m, 20H), 1.47 (tt, J = 7.4, 7.4 Hz, 2H), 2.22 (td, J = 7.4, 1.1 Hz, 2H), 3.85 (dd, J = 11.7, 5.7 Hz, 1H), 4.16 (dd, J = 11.7, 4.3 Hz, 1H), 4.42–4.49 (m, 1H), 5.03–5.07 (m, 1H), 6.72 (d, J = 7.4 Hz, 1H), 7.46 (dd, J = 8.0, 8.0 Hz, 2H), 7.54 (t, J = 8.0 Hz, 1H), 7.81 (d, J = 8.0 Hz, 2H); ^{13}C NMR (125 MHz, $CDCl_3$) δ 14.1, 18.8, 22.7, 28.2, 28.8, 29.1, 29.3, 29.4, 29.6 (3C), 29.7, 31.9, 49.2, 55.8, 61.9, 76.0, 89.9, 127.1, 128.7 (2C), 131.9 (2C), 133.8, 167.8; HRMS (FAB) calcd for $C_{25}H_{39}ClNO_2$ $[M+H]^+$, 420.2669, found: 420.2671.

3.1.11 N-[(2S,3S)-3-Chloro-1-hydroxyoctadec-4-yn-2-yl] benzamide (anti-12)

By a procedure identical with that described for synthesis of *syn*-**8** from *syn*-**7**, the propargyl chloride *anti*-**9** (100 mg, 0.22 mmol) was converted into *anti*-**12** as a white waxy solid (63.8 mg, 69% yield): mp 50–51°C; $[\alpha]_D^{23}$ –13.2 (*c* 1.23, $CHCl_3$); IR (neat): 3335 (OH), 2236 (C≡C), 1645 (C=O); ^1H NMR (500 MHz, $CDCl_3$) δ 0.88 (t, J = 6.9 Hz, 3H), 1.20–1.31 (m, 18H), 1.33–1.40 (m, 2H), 1.52 (tt, J = 7.0, 7.0 Hz, 2H), 2.27 (td, J = 7.0, 1.7 Hz, 2H), 2.44–2.60 (m, 1H), 3.90 (dd, J = 11.5, 5.2 Hz, 1H), 4.11 (dd, J = 11.5, 5.2 Hz, 1H), 4.50–4.56 (m, 1H), 5.05 (dt, J = 4.5, 1.7 Hz, 1H), 6.72 (d, J = 8.0 Hz, 1H), 7.46 (dd, J = 7.4, 7.4 Hz, 2H), 7.53 (t, J = 7.4 Hz, 1H), 7.81 (d, J = 7.4 Hz, 2H); ^{13}C NMR (125 MHz, $CDCl_3$) δ 14.1, 18.8, 22.7, 28.3, 28.9, 29.1, 29.3, 29.5, 29.6 (3C), 29.7, 31.9, 49.9, 56.2, 62.2, 75.0, 90.7, 127.1, 128.7 (2C), 131.9 (2C), 133.9, 168.0; HRMS (FAB) calcd for $C_{25}H_{39}ClNO_2$ $[M+H]^+$, 420.2669, found: 420.2670.

3.1.12 [(4S,5R)-5-(Pentadec-1-yn-1-yl)-2-phenyl-4,5-dihydrooxazol-4-yl]methanol (cis-13) and [(2R,3R)-2-(Hydroxymethyl)-3-(pentadec-1-yn-1-yl)aziridin-1-yl](phenyl)-methanone (14)

To a stirred solution of *syn*-**12** (30 mg, 0.072 mmol) in DMF (0.6 mL) was added NaH (3.2 mg, 0.079 mmol) at 0°C, and the mixture was stirred for 2.0 h at room temperature, followed by quenching with H_2O. The whole was extracted with EtOAc. The extract was washed with H_2O and brine, dried over Na_2SO_4, and concentrated under pressure to give an oily residue, which was purified by PTLC with *n*-hexane–EtOAc (1:1) to give oxazoline *cis*-**13** as a white solid (1.8 mg, 7% yield) and aziridine **14** as a colorless oil (19.1 mg, 69% yield).

cis-**13**: mp 51–52°C; $[\alpha]_D^{23}$ −98.6 (*c* 1.18, $CHCl_3$); IR (neat): 3288 (OH), 2241 (C≡C), 1649 (C=N); ^{1}H NMR (500 MHz, $CDCl_3$) δ 0.88 (t, J = 6.9 Hz, 3H), 1.20–1.32 (m, 18H), 1.34–1.42 (m, 2H), 1.54 (tt, J = 6.9, 6.9 Hz, 2H), 2.22 (t, J = 6.9 Hz, 1H), 2.28 (td, J = 7.2, 2.1 Hz, 2H), 3.90–3.96 (m, 2H), 4.45 (ddd, J = 9.5, 4.7, 4.7 Hz, 1H), 5.41 (dt, J = 9.5, 2.1 Hz, 1H), 7.41 (dd, J = 7.4, 7.4 Hz, 2H), 7.49 (t, J = 7.4 Hz, 1H), 7.96 (d, J = 7.4 Hz, 2H); ^{13}C NMR (125 MHz, $CDCl_3$) δ 14.1, 18.8, 22.7, 28.4, 28.9, 29.1, 29.3, 29.5, 29.6 (3C), 29.7, 31.9, 63.5, 70.3, 71.3, 73.8, 91.3, 127.1, 128.3 (2C), 128.4 (2C), 131.7, 163.9; HRMS (FAB) calcd for $C_{25}H_{38}NO_2$ $[M+H]^+$, 384.2903, found: 384.2903.

14: $[\alpha]_D^{27}$ −133.9 (*c* 1.84, $CHCl_3$); IR (neat): 3437 (OH), 2244 (C≡C), 1683 (C=O); ^{1}H NMR (500 MHz, $CDCl_3$) δ 0.88 (t, J = 7.0 Hz, 3H), 1.20–1.37 (m, 20H), 1.46 (tt, J = 7.0, 7.0 Hz, 2H), 2.19 (td, J = 7.0, 1.7 Hz, 2H), 2.32–2.40 (m, 1H), 3.02 (ddd, J = 6.3, 6.3, 5.9 Hz, 1H), 3.19 (dt, J = 5.9, 1.7 Hz, 1H), 3.97–4.02 (m, 2H), 7.47 (dd, J = 7.7, 7.7 Hz, 2H), 7.57 (t, J = 7.7 Hz, 1H), 8.08 (d, J = 7.7 Hz, 2H); ^{13}C NMR (125 MHz, $CDCl_3$) δ 14.1, 18.7, 22.7, 28.8, 29.1, 29.3, 29.5 (2C), 29.6 (3C), 29.7, 31.3, 31.9, 42.3, 61.7, 73.8, 85.7, 128.5, 129.4 (2C), 132.4 (2C), 133.1, 177.8; HRMS (FAB) calcd for $C_{25}H_{38}NO_2$ $[M+H]^+$, 384.2903, found: 384.2907.

3.1.13 [(4S,5S)-5-(Pentadec-1-ynyl)-2-phenyl-4,5-dihydrooxazol-4-yl]methanol (trans-13)

By a procedure identical with that described for synthesis of *cis*-**13** from *syn*-**12**, the propargyl chloride *anti*-**12** (10 mg, 0.024 mmol) was converted into *trans*-**13** as a white solid (3.3 mg, 36% yield): mp 110–111°C; $[\alpha]_D^{24}$ +3.58 (*c* 0.52, $CHCl_3$); IR (neat): 3228 (OH), 2238 (C≡C), 1650 (C=N); ^{1}H NMR (500 MHz, $CDCl_3$) δ 0.88 (t, J = 6.9 Hz, 3H), 1.23–1.32 (m, 18H), 1.33–1.40 (m, 2H), 1.52 (tt, J = 7.0, 7.0 Hz, 2H), 2.24 (td, J = 7.0, 1.5 Hz, 2H), 2.45–2.55 (m, 1H), 3.68–3.76 (m, 1H), 4.04 (d, J = 11.5 Hz, 1H), 4.35 (ddd, J = 8.0, 4.0, 3.4 Hz, 1H), 5.17

(d, $J = 8.0$ Hz, 1H), 7.37 (dd, $J = 7.4, 7.4$ Hz, 2H), 7.47 (t, $J = 7.4$ Hz, 1H), 7.89 (d, $J = 7.4$ Hz, 2H); ^{13}C NMR (125 MHz, $CDCl_3$) δ 14.1, 18.8, 22.7, 28.3, 28.9, 29.1, 29.3, 29.5, 29.6 (3C), 29.7, 31.9, 62.5, 71.0, 75.8, 76.9, 89.2, 126.8, 128.4 (2C), 130.0 (2C), 131.7, 164.7; HRMS (FAB) calcd for $C_{25}H_{38}NO_2$ $[M+H]^+$, 384.2903, found: 384.2903.

3.1.14 Synthesis of the Authentic Sample of cis-13 from the Known Compound syn-6

To a stirred solution of *syn*-**6** (40 mg, 0.091 mmol) in CH_2Cl_2 (0.7 mL) were added Et_3N (0.020 mL, 0.14 mmol), DMAP (16.7 mg, 0.14 mmol) and TsCl (21.8 mg, 0.11 mmol) at 0°C, and the mixture was stirred for 2.0 h at room temperature, followed by quenching with saturated NH_4Cl. The whole was extracted with EtOAc. The extract was washed with H_2O and brine, dried over Na_2SO_4, and concentrated under pressure to give an oily residue, which was purified by column chromatography over silica gel with *n*-hexane–EtOAc (10:1) to give the corresponding tosylate as a colorless oil. To a stirred solution of the tosylate in MeOH (0.22 mL) at 0°C was added trifluoroacetic acid (0.73 mL), and the mixture was stirred for 1.5 h at 50°C. The mixture was concentrated under reduced pressure, and the residue was dissolved in CH_2Cl_2 (0.73 mL). The solution was made neutral with $(i\text{-Pr})_2NEt$ at 0°C. Further $(i\text{-Pr})_2NEt$ (0.044 mL, 0.25 mmol) and BzCl (0.0092 mL, 0.079 mmol) were added to stirred mixture at 0°C. The mixture was stirred for 2.0 h at room temperature, followed by quenching with H_2O. The whole was extracted with EtOAc. The extract was washed successively with 1 N HCl, H_2O and brine, and dried over Na_2SO_4. The filtrate was concentrated under reduced pressure to give an oily residue, which was purified by flash chromatography over silica gel with *n*-hexane–EtOAc (3:2) to give *cis*-**13** as a white solid (14.2 mg, 40% yield). All the spectral data were in agreement with those of *cis*-**13** obtained from *syn*-**12**.

3.1.15 Synthesis of the Authentic Sample of trans-13 from the Known Compound anti-6

By a procedure identical with that described for synthesis of *cis*-**13** from *syn*-**6**, the propargyl alcohol *anti*-**6** (60 mg, 0.14 mmol) was converted into *trans*-**13** as a white solid (27 mg, 51% yield). All the spectral data were in agreement with those of *trans*-**13** obtained from *anti*-**12**.

3.1.16 General Procedure for Palladium-Catalyzed Cascade Cyclization of Propargyl Chlorides: Synthesis of (3aS,6aS,E)-2-Phenyl-6-tetradecylidene-3a,4,6,6a-tetra-hydrofuro[3,4-d]oxazole ((E)-2) from syn-12

To a stirred mixture of *syn*-**12** (40 mg, 0.095 mmol) in THF/MeOH (1.0 mL, 10:1) were added $Pd(PPh_3)_4$ (11.0 mg, 0.0095 mmol) and Cs_2CO_3 (37.1 mg, 0.114 mmol) at room temperature under argon. The mixture was stirred for 1.0 h at 50°C, and filtrated through a short pad of SiO_2 with EtOAc. The filtrate was concentrated under reduced pressure to give a yellow oil, which was purified by flash chromatography over silica gel with *n*-hexane–EtOAc (4:1) to give (*E*)-**2** as a white solid (32.4 mg, 89% yield): mp 79–80°C; $[\alpha]_D^{24}$ + 253.32 (*c* 1.38, $CHCl_3$); IR (neat): 1647 (C=N); ^1H NMR (500 MHz, $CDCl_3$) δ 0.88 (t, $J = 6.9$ Hz, 3H), 1.20–1.47 (m, 22H), 2.13–2.26 (m, 2H), 4.19 (dd, $J = 9.5$, 6.3 Hz, 1H), 4.27 (dd, $J = 9.5$, 1.7 Hz, 1H), 4.91 (ddd, $J = 8.0$, 6.3, 1.7 Hz, 1H), 5.09 (t, $J = 8.0$ Hz, 1H), 5.58 (d, $J = 8.0$ Hz, 1H), 7.40 (dd, $J = 7.4$, 7.4 Hz, 2H), 7.48 (t, $J = 7.4$ Hz, 1H), 7.93 (d, $J = 7.4$ Hz, 2H); ^{13}C NMR (125 MHz, $CDCl_3$) δ 14.1, 22.7, 27.0, 29.2, 29.4, 29.6, 29.7 (5C), 30.4, 31.9, 70.6, 74.9, 79.0, 105.0, 127.2, 128.3 (2C), 128.5 (2C), 131.6, 154.2, 164.2. *Anal.* Calcd for $C_{25}H_{37}NO_2$: C, 78.28; H, 9.72; N, 3.65. Found: C, 77.99; H, 9.80; N, 3.67 (Table 3.2, Entry 6).

3.1.17 General Procedure for Palladium-Catalyzed Cascade Cyclization of Propargyl Carbonates: Synthesis of (3aS,6aS,E)-2-Phenyl-6-tetradecylidene-3a,4,6,6a-tetra-hydrofuro[3,4-d]oxazole ((E)-2) from syn-8

To a stirred mixture of *syn*-**8** (40 mg, 0.087 mmol) in THF (0.9 mL) was added $Pd(PPh_3)_4$ (5.03 mg, 0.0044 mmol) at room temperature under argon. After stirring for 2.0 h at 50°C, concentration under reduced pressure gave a yellow oil, which was purified by flash chromatography over silica gel with *n*-hexane–EtOAc (4:1) to give (*E*)-**2** as a white solid (23.1 mg, 69% yield) (Table 3.3, Entry 2).

3.1.18 (3aS,6aS,Z)-2-Phenyl-6-tetradecylidene-3a,4,6,6a-tetrahydrofuro[3,4-d]oxazole ((Z)-2)

To a stirred mixture of *syn*-**12** (40 mg, 0.095 mmol) in THF/MeOH (1.0 mL, 10:1) was added $Pd(PPh_3)_4$ (5.03 mg, 0.0048 mmol) at room temperature under argon. After stirring for 1.5 h at 50°C, concentration under reduced pressure gave a yellow oil, which was purified by flash chromatography over silica gel with *n*-hexane–EtOAc (4:1) to give an isomeric mixture **2** (*E*:*Z* = 54:46) as a white solid (7.7 mg, 21% yield). This mixture was separated by PTLC with hexane–EtOAc (2:1) to give (*Z*)-**2** in a pure form: pale yellow oil; $[\alpha]_D^{24}$ + 143.68 (*c* 0.36, $CHCl_3$); IR (neat): 1646 (C = N); ^{1}H NMR (500 MHz, $CDCl_3$) δ 0.88 (t, $J = 6.9$ Hz, 3H), 1.20–1.40 (m, 22H), 2.02–2.20 (m, 2H), 4.25 (dd, $J = 9.5$, 6.3 Hz, 1H), 4.32 (dd, $J = 9.5$, 2.0 Hz, 1H), 4.81 (t, $J = 7.2$ Hz, 1H), 4.89 (ddd, $J = 8.6$, 6.3, 2.0 Hz, 1H), 5.37 (d, $J = 8.6$ Hz, 1H), 7.41 (dd, $J = 7.4$, 7.4 Hz, 2H), 7.48 (t, $J = 7.4$ Hz, 1H), 7.94 (d, $J = 7.4$ Hz, 2H); ^{13}C NMR (125 MHz, $CDCl_3$) δ 14.1, 22.7, 25.4, 29.3, 29.5 (2C), 29.6 (5C), 29.7, 31.9, 70.2, 75.9, 82.3, 106.0, 127.3, 128.3 (2C), 128.5 (2C), 131.5, 153.7, 164.3; HRMS (FAB) calcd for $C_{25}H_{38}NO_2$ $[M+H]^+$, 384.2903, found: 384.2900 (Table 3.2, Entry 3).

3.1.19 (3aS,6S,6aS)-2-Phenyl-6-tetradecyl-3a,4,6,6a-tetrahydrofuro[3,4-d]oxazole (15)

A mixture of (*E*)-**2** (50.0 mg, 0.13 mmol) and $(Ph_3P)_3RhCl$ (12.1 mg, 0.013 mmol) in C_6H_6/EtOH (1.3 mL, 1:1) was stirred for 25 h at 50°C under H_2, and then filtrated through a short pad of SiO_2 with EtOAc. The filtrate was concentrated under reduced pressure to give a brown solid, which was purified by flash chromatography over silica gel with *n*-hexane–EtOAc (3:1) to give **15** as a white solid (41.1 mg, 82% yield): $[\alpha]_D^{24}$ + 65.4 (*c* 1.38, $CHCl_3$) [lit $[\alpha]_D^{24}$ + 60.9 (*c* 1.05, $CHCl_3$)]. All the spectral data were in agreement with those of compound **16** in Chap. 2 (Table 3.4, Entry 6).

3.1.20 (2S,3S,4S)-4-Amino-2-tetradecyltetrahydrofuran-3-ol [Pachastrissamine (1)]

To a stirred mixture of **15** (20 mg, 0.052 mmol) in CH_2Cl_2 (0.2 mL) was added aqueous 20% H_2SO_4 (1.0 mL), and the mixture was stirred at 120°C for 43 h in a seal tube. The mixture was quenched by addition of 10 N NaOH at 0°C, and the whole was extracted with $CHCl_3$. The extract was dried over Na_2SO_4. The filtrate was concentrated under reduced pressure to give a white solid, which was purified by flash chromatography over silica gel with $CHCl_3$–MeOH–28% NH_4OH (95:4:1) to give **1** as a white solid (12.5 mg, 80% yield): mp 97–98°C; $[\alpha]_D^{25}$ + 18.9 (*c* 0.77, EtOH) [lit $[\alpha]_D$+18 (*c* 0.1, EtOH)]; IR (neat): 3341 (OH and NH); ^{1}H NMR (500 MHz, $CDCl_3$) δ 0.88 (t, $J = 7.0$ Hz, 3H), 1.20–1.49 (m, 24H), 1.59–1.73 (m, 2H), 1.80–2.20 (m, 2H), 3.51 (dd, $J = 8.3$, 6.9 Hz, 1H), 3.60–3.70 (m, 1H), 3.73 (ddd, $J = 7.7$, 6.6, 3.9 Hz, 1H), 3.87 (dd, $J = 4.7$, 3.9 Hz, 1H), 3.92 (dd, $J = 8.3$, 7.4 Hz, 1H); ^{13}C NMR (125 MHz, $CDCl_3$) δ 14.1, 22.7, 26.3, 29.3,

29.4, 29.6 (6C), 29.7, 29.8, 31.9, 54.3, 71.8, 72.4, 83.2. *Anal.* Calcd for $C_{18}H_{37}NO_2$: C, 72.19; H, 12.45; N, 4.68. Found: C, 71.79; H, 12.14; N, 4.57.

3.1.21 (2S,3S,4S)-4-(Benzylamino)-2-tetradecyltetrahydrofuran-3-ol (N-Benzylpachastrissamine) (16)

To a stirred solution of **15** (120 mg, 0.31 mmol) in CH_2Cl_2 (6.0 mL) was added DIBAL-H in toluene (1.01 M; 1.24 mL, 1.24 mmol) at 0°C. After stirring for 20 min at this temperature, the mixture was allowed to warm to room temperature. The mixture was stirred for 2.0 h at this temperature and quenched with 2 N Rochelle salt. After stirring for 3.0 h, the whole was extracted with EtOAc. The extract was washed with brine, and dried over Na_2SO_4. Concentration under reduced pressure followed by flash chromatography over silica gel with $CHCl_3$–MeOH (15:1) gave **16** as a white solid: mp 72–73°C; $[\alpha]_D^{25}$ + 14.5 (*c* 0.99, $CHCl_3$); IR (neat): 3334 (NH and OH); ^{1}H NMR (500 MHz, $CDCl_3$) δ 0.88 (t, J = 6.9 Hz, 3H), 1.22–1.43 (m, 24H), 1.62–1.75 (m, 2H), 2.39–2.90 (m, 2H), 3.44 (ddd, J = 7.4, 7.4, 4.6 Hz, 1H), 3.55 (dd, J = 8.6, 7.4 Hz, 1H), 3.70 (ddd, J = 6.9, 6.9, 2.9 Hz, 1H), 3.77 (d, J = 13.2 Hz, 1H), 3.84 (d, J = 13.2 Hz, 1H), 3.89–3.92 (m, 1H), 3.91–3.94 (m, 1H), 7.26–7.37 (m, 5H); ^{13}C NMR (125 MHz, $CDCl_3$) δ 14.1, 22.7, 26.3, 29.3, 29.4, 29.6 (5C), 29.7 (2C), 29.8, 31.9, 52.7, 61.2, 69.7, 70.4, 83.5, 127.5, 128.1 (2C), 128.6 (2C), 139.2; HRMS (FAB) calcd for $C_{25}H_{44}NO_2$ $[M+H]^+$, 390.3372, found: 390.3372.

3.1.22 (2S,3S,4S)-4-Amino-2-tetradecyltetrahydrofuran-3-ol (Pachastrissamine) (1) from 16

A mixture of **16** (20.0 mg, 0.051 mmol) and 20% w/w $Pd(OH)_2/C$ (3.6 mg, 0.0051 mmol) in EtOAc (0.8 mL) was stirred at 50°C under H_2. After stirring for 12 h, further EtOAc (0.4 mL) was added to stirred mixture. The mixture was stirred for 9 h at 50°C, and filtrated through a short pad of Celite with EtOAc. The filtrate was concentrated under reduced pressure to give a white solid, which was purified by flash chromatography over silica gel with $CHCl_3$–MeOH–28% NH_4OH (95:4:1) to give **1** as a white solid (13.1 mg, 86% yield).

References

1. Herold P (1988) Helv Chim Acta 71:354–362
2. De Jonghe S, Van Overmeire I, Van Calenbergh S, Hendrix C, Busson R, De Keukeleire D, Herdewijin P (2000) Eur J Org Chem 3177–3183

3. Roush DM, Patel MM (1985) Synth Commun 15:675–679
4. Freedman J, Vaal MJ, Huber EW (1991) J Org Chem 56:670–672
5. Lipshutz BH, Miller TA (1990) Tetrahedron Lett 31:5253–5256
6. Okuda M, Tomioka K (1994) Tetrahedron Lett 35:4585–4586
7. Ohno H, Okano A, Kosaka S, Tsukamoto K, Ohata M, Ishihara K, Maeda H, Tanaka T, Fujii N (2008) Org Lett 10:1171–1174
8. Okano A, Tsukamoto K, Kosaka S, Maeda H, Oishi S, Tanaka T, Fujii N, Ohno H (2010) Chem Eur J 16:8410–8418
9. Okano A, Oishi S, Tanaka T, Fujii N, Ohno H (2010) J Org Chem 75:3396–3400
10. Ohno H, Toda A, Takemoto Y, Fujii N, Ibuka T (1999) J Chem Soc Perkin Trans 1:2949–2962
11. Ohno H, Hamaguchi H, Ohata M, Tanaka T (2003) Angew Chem Int Ed 42:1749–1753
12. Ohno H, Hamaguchi H, Ohata M, Kosaka S, Tanaka T (2004) J Am Chem Soc 126:8744–8754
13. Hamaguchi H, Kosaka S, Ohno H, Tanaka T (2005) Angew Chem Int Ed 44:1513–1517
14. Hamaguchi H, Kosaka S, Ohno H, Fujii N, Tanaka T (2007) Chem Eur J 13:1692–1708
15. Ogoshi S, Tsutsumi K, Nishiguchi S, Kurosawa H (1995) J Organomet Chem 493:C19–C21
16. Tsutsumi K, Ogoshi S, Nishiguchi S, Kurosawa H (1998) J Am Chem Soc 120:1938–1939
17. Tsutsumi K, Kawase T, Kakiuchi K, Ogoshi S, Okada Y, Kurosawa H (1999) Bull Chem Soc Jpn 72:2687–2692
18. Casey CP, Nash JR, Yi CS, Selmeczy AD, Chung S, Powell DR, Hayashi RK (1998) J Am Chem Soc 120:722–733
19. Ogoshi S, Kurosawa H (2003) J Synth Org Chem Jpn 61:14–23
20. Labrosse J-R, Lhoste P, Delbecq F, Sinou D (2003) Eur J Org Chem 2813–2822
21. Yoshida M, Fujita M, Ihara M (2003) Org Lett 5:3325–3327
22. Mikami K, Yoshida A (1997) Angew Chem Int Ed Engl 36:858–860
23. Ogoshi S, Nishida T, Shinagawa T, Kurosawa H (2001) J Am Chem Soc 123:7164–7165
24. Yoshida M, Gotou T, Ihara M (2004) Tetrahedron Lett 45:5573–5575
25. Molander GA, Sommers EM, Baker SR (2006) J Org Chem 71:1563–1568
26. Yoshida M, Hayashi M, Shishido K (2007) Org Lett 9:1643–1646
27. Yoshida M, Okada T, Shishido K (2007) Tetrahedron 63:6996–7002
28. Vaz B, Pereira R, Pérez M, Ávarez R, De Lera AR (2008) J Org Chem 73:6534–6541
29. Crabtree RH, Felkin H, Morris GE (1977) J Organomet Chem 141:205–215
30. Crabtree RH, Davis MW (1986) J Org Chem 51:2655–2661
31. Lehr P, Billich A, Charpiot B, Ettmayer P, Scholz D, Rosenwirth B, Gstach H (1996) J Med Chem 39:2060–2067
32. Nordin IC (1966) J Heterocycl Chem 3:531–532
33. Wilson SR, Mao DT, Khatri HN (1980) Synth Commun 10:17–23
34. Meyers AI, Himmelsbach RJ, Reuman M (1983) J Org Chem 48:4053–4058
35. Nagamitsu T, Sunazuka T, Tanaka H, Omura S, Sprengeler PA, Smith AB III (1996) J Am Chem Soc 118:3584–3590
36. Feldman KS, Cutarelli TD, Florio RD (2002) J Org Chem 67:8528–8537

Part II
Total Synthesis of Lysergic Acid, Lysergol, and Isolysergol

Chapter 4
Palladium-Catalyzed Domino Cyclization of Amino Allenes Bearing a Bromoindolyl Group and Its Application to Total Synthesis of Ergot Alkaloids

Abstract Ergot alkaloids and their synthetic analogs have been reported to exhibit broad biological activity. The author investigated direct construction of the C/D ring system of ergot alkaloids based on palladium-catalyzed domino cyclization of amino allenes. With this biscyclization as the key step, total synthesis of (±)-lysergic acid, (±)-lysergol and (±)-isolysergol was achieved.

Ergot alkaloids are pharmacologically important indole alkaloids produced by the fungus *Claviceps purpurea*, which grows parasitically on rye and other grains (for isolation of lysergic acid, see Refs. [1, 2]; for isolation of lysergol, see Ref. [3]; for isolation of isolysergol, see Ref. [4]). These alkaloids have been reported to exhibit broad biological activity, and several synthetic derivatives such as pergolide or bromocriptine are also used as anti-prolactin and anti-Parkinson's disease drugs [5, 6]. The characteristic structural feature of these alkaloids is a [*cd*]-fused indole, which contains the $\Delta^{9,10}$-double bond and chiral centers at C5 and C8 (Fig. 4.1). Owing to their biological importance as well as structural appeal, ergot alkaloids, particularly lysergic acid (**1**), have been the target of many synthetic studies, but most of the previous syntheses relied on a stepwise linear approach for construction of the C/D ring system (for synthesis of lysergic acid, see Refs. [7–16]; for synthesis of lysergol and isolysergol, see Refs. [17–21]). For one exception is Oppolzer's strategy, which is based on simultaneous construction of C/D rings by an intramolecular imino-Diels–Alder reaction [10].

The author expected palladium-catalyzed domino cyclization of amino allenes of the type **5** bearing a protected 4-bromoindol-3-yl group (Scheme 4.1) to provide direct access to the core structure of ergot alkaloids **4**, including lysergic acid (**1**), lysergol (**2**), and isolysergol (**3**). The challenges in this domino cyclization are sequential regioselective formation of a carbon–carbon bond and a carbon–nitrogen bond for the construction of the desired 6,6-fused C/D ring system.

Retrosynthetic analysis of the amino allenes **5** is shown in Scheme 4.1. The author envisioned that the allene unit of **5** can be constructed by Claisen

S. Inuki, *Total Synthesis of Bioactive Natural Products by Palladium-Catalyzed Domino Cyclization of Allenes and Related Compounds*, Springer Theses,
DOI: 10.1007/978-4-431-54043-4_4, © Springer 2012

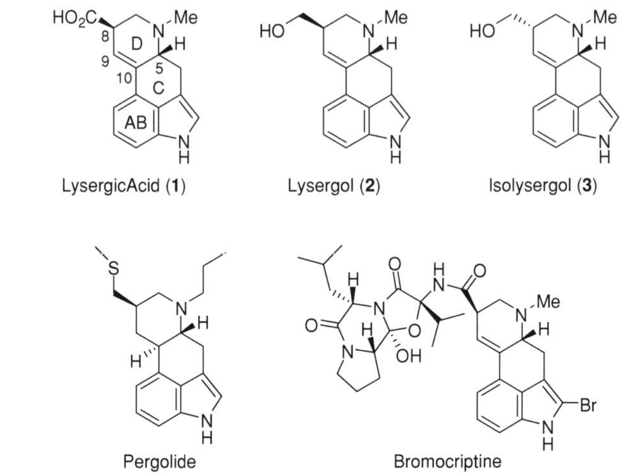

Fig. 4.1 Indole alkaloids of the ergot family and synthetic derivatives

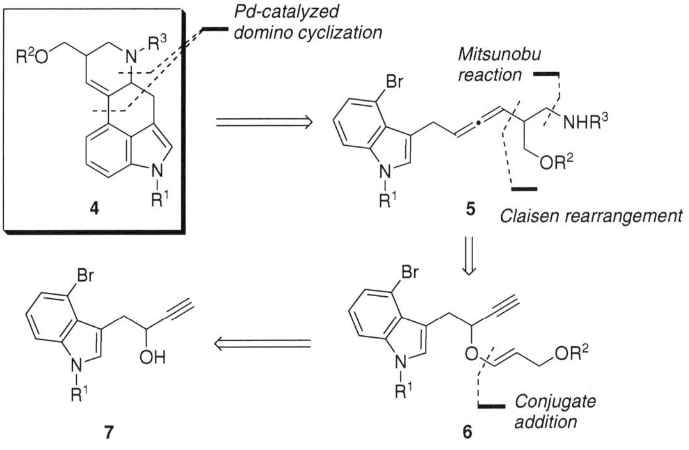

Scheme 4.1 Retrosynthetic analysis of **4**

rearrangement of enol ether **6**, which could be readily obtained by conjugate addition of propargyl alcohol **7** to methyl propiolate followed by reduction/protection.

Preparation of the requisite enol ether of the type **6** for Claisen rearrangement is outlined in Scheme 4.2. 3-(Bromomethyl)indole **10** is easily accessible from commercially available 4-bromoindole **8** [22]. Lithiation and addition of 1,3-dithiane **11** [23, 24] to the bromide **10** gave thioacetal **12** in 96% yield. Subsequent functional-group modifications, including hydrolysis of the thioacetal [25], reduction, desilylation, and conjugate addition to methyl propiolate, provided

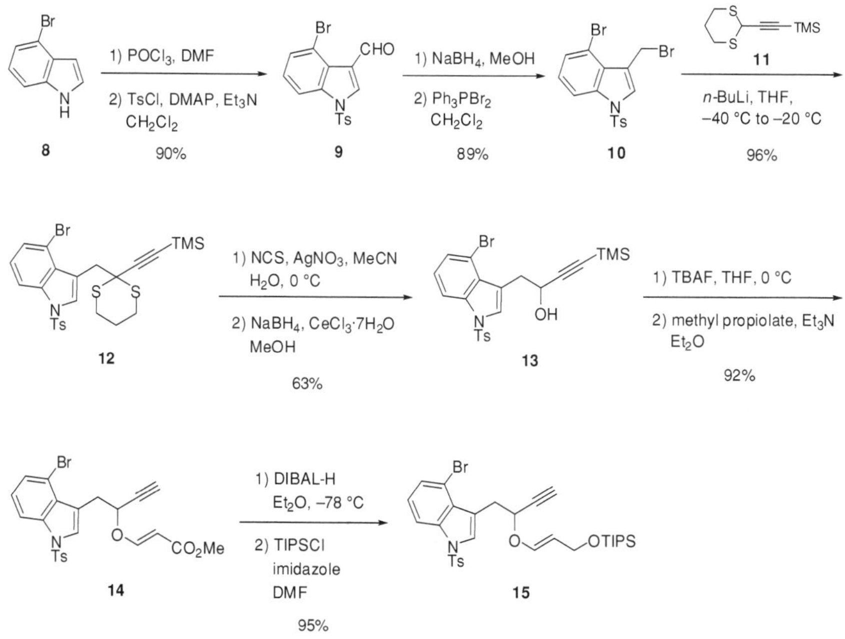

Scheme 4.2 Synthesis of propargyl ether **15**

the enoate **14** [26]. The propargyl vinyl ether **15** was obtained by DIBAL reduction and silylation of **14**. Claisen rearrangement under thermal conditions (*m*-xylene, 170 °C) gave the desired allenic alcohol **16** (**a**:**b** = ca. 33:67) in only 38% yield (Table 4.1, entry 1). Microwave irradiation [27] in $CHCl_3$ dramatically improved the yield to 82% (entry 2).[1] Furthermore, use of 5 mol % of gold-oxo complex $[(Ph_3PAu)_3O]BF_4$ resulted in 78% yield of **16**, in favor of the opposite diastereomer (**a**:**b** = ca. 80:20, entry 3) [29]. Mitsunobu reaction of **16** with $NsNH_2$ or TsNHFmoc [30] (followed by piperidine treatment) gave *N*-nosyl and *N*-tosylamide derivatives **18** and **19** (**a**:**b** = 80:20), respectively (Scheme 4.3).

[1] The relative configuration of **16a** was determined by NOE analyses of the corresponding pyran, obtained by Au-catalyzed stereospecific cyclization [28].

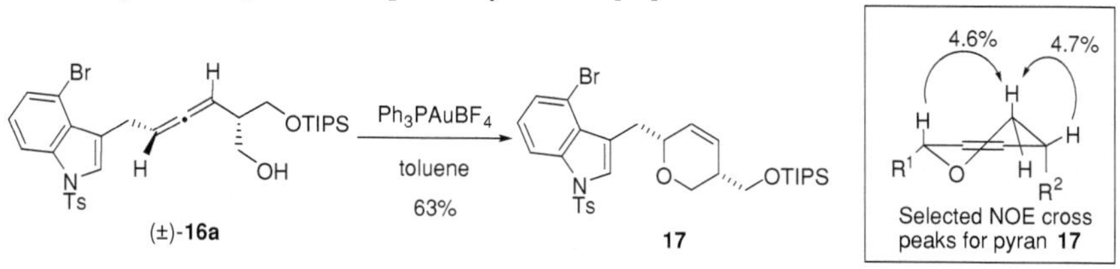

(±)-**16a** → Ph_3PAuBF_4, toluene, 63% → **17**

Selected NOE cross peaks for pyran **17** (4.6%, 4.7%)

Table 4.1 Claisen rearrangement of propargyl ether **15**

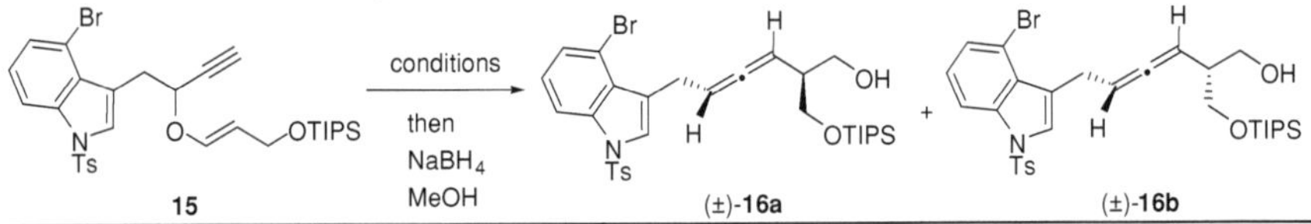

Entry	Conditions[a]	Yield (%)[b]	dr (**a**:**b**)[c]
1	*m*-xylene, 170 °C, 50 min	38	ca. 33:67
2	$CHCl_3$, MW, 120 °C, 12 min, 150 °C, 12 min	82	ca. 33:67
3	$[(Ph_3PAu)_3O]BF_4$ (5 mol %), CH_2Cl_2, 40 °C, 10 h	78	ca. 80:20

[a] MW = microwave irradiation
[b] Isolated yields after reduction with $NaBH_4$
[c] Determined by HPLC and ^{1}H NMR analysis

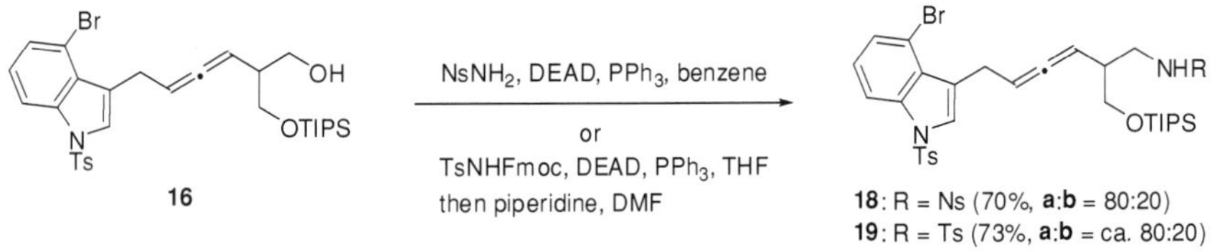

Scheme 4.3 Synthesis of allenic amides **18** and **19**

The author next investigated construction of the ergot alkaloid skeleton via the palladium-catalyzed domino cyclization (Table 4.2). The reaction was conducted using a 80:20 diastereomixture of **18** and **19** because separating the diastereomeric mixtures resulting from Claisen rearrangement was difficult. Reaction of **18** with 5 mol % of $Pd(PPh_3)_4$ and Na_2CO_3 in DMF at 100 °C afforded desired product **20** in 31% yield (**a**:**b** = 84:16, entry 1). Among the several bases investigated, K_2CO_3 has proven to be the most effective to give 83% of **20** as a 73:27 diastereomixture (entry 3).[2] Although the reaction at 120 °C slightly decreased the yield of the desired product, unidentified side products were easily removed from the desired product **20** (entry 4). Changing the solvent from DMF to toluene, dioxane or DMSO did not enhance the yield of desired product (entries 5–7). Further screening using $Pd(OAc)_2/PPh_3$ (entry 8), $PdCl_2$(dppf) (entry 9), $Pd(OAc)_2/P(o\text{-}tol)_3$ (entry 10) and $Pd(OAc)_2$/*rac*-BINAP (entry 11) was done. As diastereoselectivity improved, yield of desired product decreased (entries 3, 8–10), except for using $Pd(OAc)_2$/*rac*-BINAP (entry 11). When the *N*-tosyl derivative **19** was

[2] The relative configurations of **20a** and **20b** were confirmed by their conversion to isolysergol and lysergol, respectively.

Table 4.2 Palladium-catalyzed domino cyclization[a]

Pd (5 mol%)/ ligand (10 mol%); base (3.0 eq.) DMF, 100 °C

(±)-**18**: R = Ns (**a**:**b** = 80:20)
(±)-**19**: R = Ts (**a**:**b** = ca. 80:20)

20a: R = Ns
21a: R = Ts

20b: R = Ns
21b: R = Ts

Entry	Pd/ligand	Solvent	Base	Yield (%)[b]	dr (**a**:**b**)[c]
1	$Pd(PPh_3)_4$	DMF	Na_2CO_3	31	84:16
2	$Pd(PPh_3)_4$	DMF	Cs_2CO_3	41	75:25
3	$Pd(PPh_3)_4$	DMF	K_2CO_3	83	73:27
4[d]	$Pd(PPh_3)_4$	DMF	K_2CO_3	78	74:26
5	$Pd(PPh_3)_4$	toluene	K_2CO_3	trace	ND[f]
6	$Pd(PPh_3)_4$	dioxane	K_2CO_3	68	80:20
7[d]	$Pd(PPh_3)_4$	DMSO	K_2CO_3	68	74:26
8	$Pd(OAc)_2/PPh_3$	DMF	K_2CO_3	61	88:12
9	$PdCl_2$(dppf)	DMF	K_2CO_3	41	92:8
10	$Pd(OAc)_2$/P(*o*-tol)$_3$	DMF	K_2CO_3	20	>95:5
11[e]	$Pd(OAc)_2$/*rac*-BINAP	DMF	K_2CO_3	31	72:28
12[g]	$Pd(PPh_3)_4$	DMF	K_2CO_3	65	87:13

[a] Reactions were carried out using a diastereomixture of **18** or **19** (**a**:**b** = 80:20) at 0.06 M for 2.5–5 h
[b] Isolated yields
[c] Determined by ^{1}H NMR analysis
[d] Reaction was performed at 120 °C
[e] Reactions were carried out using Pd (5 mol%) and ligand (5 mol%)
[f] Not determined
[g] Reaction was carried out using a substrate **19** at 120 °C

employed, desired product **21** was isolated in 65% yield with good diastereoselectivity (87:13, entry 12).[3]

To obtain some mechanistic insight of the domino cyclization, diastereomerically pure **18a** and **18b** (obtained by careful HPLC separation of **16** followed by Mitsunobu reaction) were subjected to the reaction conditions shown in entry 4 (Table 4.2). Domino cyclization of the major isomer **18a** gave an 83:17

[3] The relative configurations of **21a** were confirmed by derivatization of **20a** to the same compound.

20a → 1) $HSCH_2CO_2H$, $LiOH{\cdot}H_2O$, DMF; 2) TsCl, Et_3N, CH_2Cl_2; 76% → **21a**

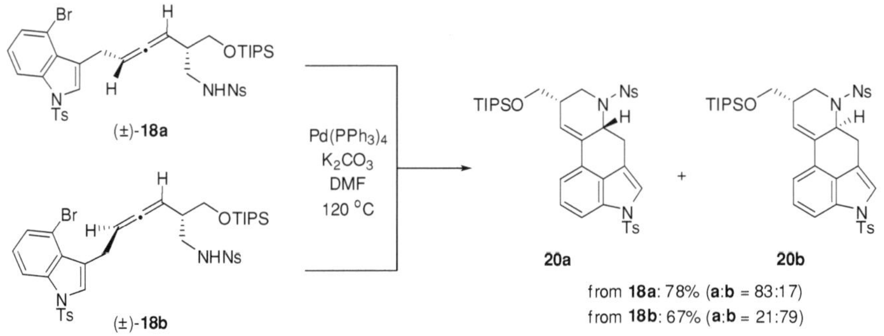

Scheme 4.4 Palladium-catalyzed domino cyclization of diastereomerically pure substrates

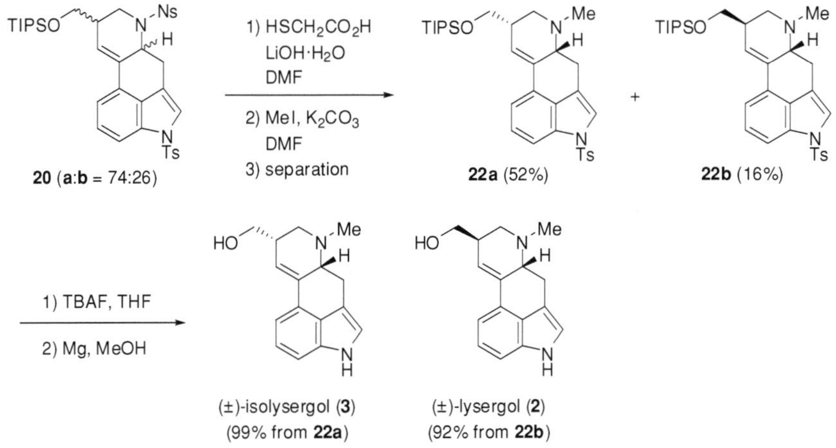

Scheme 4.5 Total synthesis of (±)-lysergol (**2**) and (±)-isolysergol (**3**)

diastereomixture, in 78% yield, in which the major cyclized product **20a** predominated (Scheme 4.4). In contrast, reaction of the minor isomer **18b** favored the diastereomer **20b** (**a**:**b** = 21:79) in 67% yield.

With the ergot alkaloid derivatives **20** and **21** bearing the requisite functionalities in hand, the final stage was set for the completion of the total synthesis of lysergic acid, lysergol and isolysergol (Schemes 4.5, 4.6). Deprotection of the Ns group of **20** and *N*-methylation gave a separable mixture of diastereomers, each of which was readily converted into (±)-isolysergol (**3**) and (±)-lysergol (**2**) by removal of TIPS and Ts groups (Scheme 4.5). The author chose tosylamide **21** as the precursor of lysergic acid (Scheme 4.6).[4] Cleavage of the TIPS group of **21**,

[4] Cleavage of nosyl group in the ester derived from **20** was less effective (20–30% yield) under standard conditions.

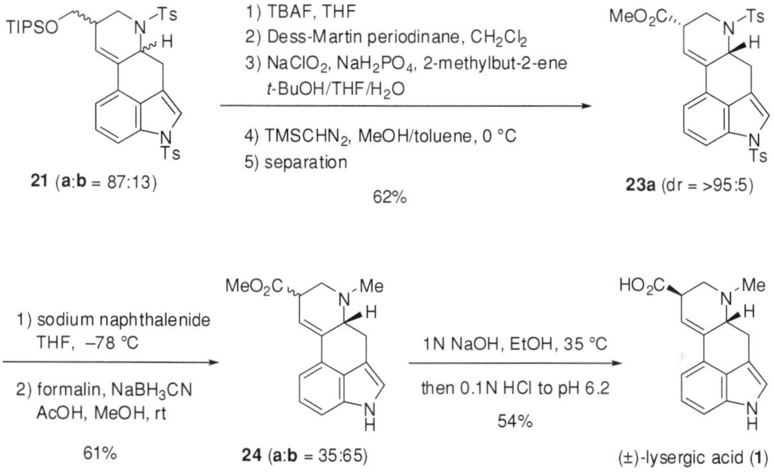

21 (**a**:**b** = 87:13)

1) TBAF, THF
2) Dess-Martin periodinane, CH_2Cl_2
3) $NaClO_2$, NaH_2PO_4, 2-methylbut-2-ene
t-BuOH/THF/H_2O
4) $TMSCHN_2$, MeOH/toluene, 0 °C
5) separation
62%

23a (dr = >95:5)

1) sodium naphthalenide
THF, −78 °C
2) formalin, $NaBH_3CN$
AcOH, MeOH, rt
61%

24 (**a**:**b** = 35:65)

1N NaOH, EtOH, 35 °C
then 0.1N HCl to pH 6.2
54%

(±)-lysergic acid (**1**)

Scheme 4.6 Total synthesis of (±)-lysergic acid (**1**)

oxidation of the resulting primary alcohol by standard protocol, and esterification with $TMSCHN_2$ gave the corresponding methyl ester **23a** (62%, 4 steps).[5] Removal of tosyl groups with sodium naphthalenide and subsequent *N*-methylation led to a diastereomixture of methyl isolysergate **24a** and lysergate **24b** (35:65). Total synthesis of (±)-lysergic acid was completed by hydrolysis of **24** with NaOH, accompanying isomerization to the favorable isomer [15]. Physical data were in agreement with those of natural and synthetic lysergic acid, lysergol and isolysergol reported in the literature [15, 16, 20].

In conclusion, the author developed a novel entry to direct construction of ergot alkaloids skeleton based on palladium-catalyzed domino cyclization of amino allenes. With this biscyclization as the key step, total synthesis of (±)-lysergic acid, (±)-lysergol and (±)-isolysergol was achieved.

---

[5] The relative configuration of **23a** was confirmed by conversion to **21a** [31].

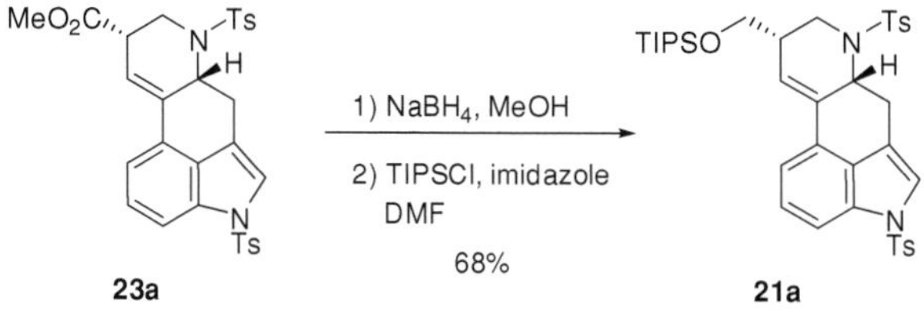

23a

1) $NaBH_4$, MeOH
2) TIPSCl, imidazole
DMF
68%

21a

4.1 Experimental Section

4.1.1 General Methods

All moisture-sensitive reaction were performed using syringe-septum cap techniques under an argon atmosphere and all glassware was dried in an oven at 80 °C for 2 h prior to use. Reactions at –78 °C employed a CO_2–MeOH bath. Melting points were measured by a hot stage melting point apparatus and are uncorrected. For flash chromatography, Wakosil C-300 or Wakogel C-300E was employed. ^{1}H NMR spectra were recording using a JEOR AL-400 or JEOL ECA-500 spectrometer, and chemical shifts are reported in δ (ppm) relative to TMS (in $CDCl_3$) as internal standard. ^{13}C NMR spectra were recorded using a JEOR AL-400 or JEOL ECA-500 spectrometer and referenced to the residual $CHCl_3$ signal. NOE spectra were recorded on 500 MHz instruments. Chemical shifts were reported in parts per million with the residual solvent peak used as an internal standard. ^{1}H NMR spectra are tabulated as follows: chemical shift, multiplicity (b = broad, s = singlet, d = doublet, t = triplet, q = quartet, m = multiplet), number of protons, and coupling constant(s). Exact mass (HRMS) spectra were recorded on a JMS-HX/HX 110A mass spectrometer. Infrared (IR) spectra were obtained on a JASCO FT/IR-4100 FT-IR spectrometer with JASCO ATR PRO410-S. Microwave reaction was conducted in a sealed glass vessel (capacity 10 mL) using CEM Discover microwave reactor. The temperature was monitored using IR sensor mounted under the reaction vessel. For analytical HPLC, a Cosmosil 5C18-ARII column (4.6 × 250 mm, Nacalai Tesque Inc., Kyoto, Japan) was employed on a Shimadzu LC-10ADvp (Shimadzu Corp., Ltd., Kyoto, Japan). Preparative HPLC was performed using a Cosmosil 5C18-ARII column (20 × 250 mm, Nacalai Tesque Inc.) on a Shimadzu LC-6AD (Shimadzu Corp., Ltd.).

4.1.2 4-Bromo-1-tosyl-1H-indole-3-carbaldehyde (9)

The formylation of 4-bromoindole was carried out according to the method of Lauchli and Shea [32]. To a stirred DMF (6 mL) was added $POCl_3$ (0.98 mL, 10.5 mmol) at 0 °C under argon. The solution was stirred for 2 min, and then 4-bromoindole **8** (940 mg, 4.7 mmol) in DMF (5 mL) was added. The mixture was stirred for 1 h at room temperature and was slowly quenched with KOH (2.66 g) in water (10 mL). The reaction mixture was left to cool overnight, and was then partitioned between EtOAc and saturated aqueous $NaHCO_3$. The organic layer was washed with brine, dried over $MgSO_4$, and concentrated under reduced pressure to give off a crude aldehyde as a white solid. To a stirred solution of this aldehyde in CH_2Cl_2 (40 mL) were added TsCl (1.08 g, 5.6 mmol), Et_3N (1.05 mL, 7.5 mmol) and DMAP (57.4 mg, 0.47 mmol) at 0 °C, and the mixture was stirred for 2 h at room temperature. The mixture was made acidic with 1N HCl, and whole was extracted with EtOAc. The extract was washed with brine, dried over $MgSO_4$, and concentrated under pressure to give a white solid, which was purified by column chromatography over silica gel with *n*-hexane–EtOAc (7:1) to give **9** (1.59 g, 90% yield). Recrystallization from *n*-hexane–chloroform gave pure **9** as colorless crystals: mp 176–177 °C; IR (neat): 1676 (C=O), 1381 (NSO_2), 1176 (NSO_2); ^1H NMR (500 MHz, $CDCl_3$) δ 2.38 (s, 3H), 7.24 (dd, J = 8.2, 8.2 Hz, 1H), 7.30 (d, J = 8.3 Hz, 2H), 7.54 (d, J = 8.2 Hz, 1H), 7.83 (d, J = 8.3 Hz, 2H), 8.00 (d, J = 8.2 Hz, 1H), 8.41 (s, 1H), 10.29 (s, 1H); ^{13}C NMR (100 MHz, $CDCl_3$) δ 21.7, 112.9, 113.9, 122.0, 126.1, 127.0, 127.3 (2C), 128.9, 130.4 (2C), 132.0, 134.1, 136.2, 146.4, 186.2. *Anal.* Calcd for $C_{16}H_{12}BrNO_3S$: C, 50.81; H, 3.20; N, 3.70. Found: C, 50.81; H, 3.16; N, 3.71.

4.1.3 4-Bromo-3-(bromomethyl)-1-tosyl-1H-indole (10)

To a stirred solution of the aldehyde **9** (4.30 g, 11.4 mmol) in MeOH (300 mL) was added $NaBH_4$ (1.24 g, 32.7 mmol) at 0 °C. After stirring for 1.5 h at room temperature, H_2O was added, and the mixture was concentrated under reduced pressure. The residue was diluted with Et_2O, and the organic phase was separated and washed with brine and dried over $MgSO_4$. The filtrate was concentrated under reduced pressure to give a crude alcohol as a white solid, which was used without further purification. To a stirred solution of this alcohol in CH_2Cl_2 (60 mL) was added Ph_3PBr_2 (5.30 g, 12.5 mmol) in CH_2Cl_2 (50 mL). The mixture was stirred overnight at room temperature. Concentration under reduced pressure gave an oily residue, which was purified by flash chromatography over silica gel with *n*-hexane–EtOAc (7:1) to give **10** (4.46 g, 89% yield). Recrystallization from *n*-hexane–chloroform gave pure **10** as colorless crystals: mp 157–158 °C; IR (neat): 1375 (NSO_2), 1173 (NSO_2); ^1H NMR (500 MHz, $CDCl_3$) δ 2.36 (s, 3H), 4.88 (s, 2H), 7.17 (dd, J = 8.1, 8.1 Hz, 1H), 7.26 (d, J = 8.0 Hz, 2H), 7.42 (d, J = 8.1 Hz, 1H), 7.75 (s, 1H), 7.76 (d, J = 8.0 Hz, 2H), 7.93 (d, J = 8.1 Hz, 1H); ^{13}C NMR (100 MHz, $CDCl_3$) δ 21.6, 24.7, 112.9, 114.2, 119.6, 126.1, 127.0 (2C), 127.1, 127.9, 128.2, 130.1 (2C), 134.7, 136.3, 145.6. *Anal.* Calcd for $C_{16}H_{13}Br_2NO_2S$: C, 43.36; H, 2.96; N, 3.16. Found: C, 43.57; H, 2.75; N, 2.90.

4.1.4 4-Bromo-1-tosyl-3-[2-(trimethylsilylethynyl) -1,3-dithian-2-yl]methyl-1H-indole (12)

To a stirred solution of the 2-(trimethylsilylethynyl)-1,3-dithiane **11** (38.3 mg, 0.177 mmol) in THF (1 mL) was added *n*-BuLi (1.65 M solution in hexane; 0.12 mL, 0.195 mmol) at –40 °C under argon. After stirring for 1 h with warming to −20 °C, a solution of the bromide **10** (72.5 mg, 0.164 mmol) in THF (0.2 mL) was added to this reagent at −20 °C. The mixture was stirred for 2 h at this temperature and quenched with H_2O. The whole was extracted with Et_2O. The extract was washed with brine and dried over $MgSO_4$. The filtrate was concentrated under reduced pressure to give an oily residue, which was purified by flash chromatography over silica gel with *n*-hexane–EtOAc (7:1) to give **12** (90.6 mg, 96% yield). Recrystallization from MeCN gave pure **12** as colorless crystals: mp 138–139 °C; IR (neat): 2157 (C≡C), 1374 (NSO_2), 1173 (NSO_2); ^{1}H NMR (500 MHz, $CDCl_3$) δ 0.05 (s, 9H), 1.82–1.92 (m, 1H), 2.14-2.21 (m, 1H), 2.33 (s, 3H), 2.83 (ddd, J = 13.9, 3.3, 3.3 Hz, 2H), 3.29–3.37 (m, 2H), 3.78 (s, 2H), 7.07 (dd, J = 8.0, 8.0 Hz, 1H), 7.18 (d, J = 8.2 Hz, 2H), 7.36 (d, J = 8.0 Hz, 1H), 7.72 (d, J = 8.2 Hz, 2H), 7.91 (d, J = 8.0 Hz, 1H), 7.94 (s, 1H); ^{13}C NMR (125 MHz, $CDCl_3$) δ 0.00 (3C), 21.8, 25.8, 29.1 (2C), 36.7, 47.1, 93.1, 103.2, 113.1, 115.0, 116.1, 125.3, 127.2 (2C), 128.4, 128.6, 129.5, 130.1 (2C), 135.2, 136.1, 145.2. *Anal.* Calcd for $C_{25}H_{28}BrNO_2S_3Si$: C, 51.89; H, 4.88; N, 2.42. Found: C, 51.66; H, 4.78; N, 2.24.

4.1.5 (±)-1-(4-Bromo-1-tosyl-1H-indol-3-yl)-4-(trimethylsilyl)but-3-yn-2-ol (13)

To a stirred mixture of NCS (786 mg, 5.89 mmol) and $AgNO_3$ (1.03 g, 6.06 mmol) in MeCN (25 mL) and H_2O (5 mL) was added thioacetal **12** (1.00 g, 1.73 mmol) in MeCN (8 mL) at 0 °C. The mixture was stirred for 5 min at this temperature and quenched with saturated Na_2SO_3, saturated $NaHCO_3$ and brine (1:1:1). The mixture was filtered through a short pad of Celite with EtOAc. The filtrate was extracted with Et_2O. The extract was washed with saturated Na_2SO_3, saturated $NaHCO_3$ and brine (1:1:1), brine and dried over $MgSO_4$. The filtrate was concentrated under reduced pressure to give a yellow oily residue, which was used without further purification. To a stirred solution of the crude ketone in MeOH (50 mL) was added $CeCl_3{\cdot}7H_2O$ (838 mg, 2.25 mmol) at room temperature. After stirring for 10 min, $NaBH_4$ (118 mg, 3.11 mmol) was added to this solution at −20 °C. The mixture was stirred for 1 h at this temperature and quenched with H_2O. The mixture was concentrated under reduced pressure. The residue was diluted with Et_2O, and the extract was washed with brine and dried over $MgSO_4$. The filtrate was concentrated under reduced pressure to give an oily residue, which was purified by flash chromatography over silica gel with *n*-hexane–EtOAc (4:1) to give **13** as a white amorphous solid (530 mg, 63% yield). IR (neat): 3540 (OH), 2172 (C≡C), 1373 (NSO_2), 1173 (NSO_2); ^{1}H NMR (500 MHz, $CDCl_3$) δ 0.16 (s, 9H), 1.90 (d, J = 5.1 Hz, 1H), 2.35 (s, 3H), 3.33 (dd, J = 14.0, 6.8 Hz, 1H), 3.42 (dd, J = 14.0, 6.8 Hz, 1H), 4.72 (ddd, J = 6.8, 6.8, 5.1 Hz, 1H), 7.11 (dd, J = 8.0, 8.0 Hz, 1H), 7.23 (d, J = 8.4 Hz, 2H), 7.37 (d, J = 8.0, 1H), 7.59 (s, 1H), 7.73 (d, J = 8.4 Hz, 2H), 7.94 (d, J = 8.0 Hz, 1H); ^{13}C NMR (125 MHz, $CDCl_3$) δ 0.00 (3C), 21.8, 34.6, 63.1, 90.8, 105.9, 113.1, 114.6, 117.8, 125.5, 127.1

(2C), 127.2, 128.1, 128.9, 130.2 (2C), 135.2, 136.5, 145.4; HRMS (FAB) calcd $C_{22}H_{23}BrNO_3SSi$: $[M–H]^-$, 488.0357; found: 488.0351.

4.1.6 *Methyl (±)-(E)-3-[1-(4-Bromo-1-tosyl-1H -indol-3-yl)but-3-yn-2-yloxy]acrylate (14)*

To a stirred solution of the alcohol **13** (84.2 mg, 0.17 mmol) in THF (3 mL) was added TBAF (1.00 M solution in THF; 0.22 mL, 0.22 mmol) at 0 °C. The mixture was stirred for 1 h at this temperature and quenched with H_2O. The whole was extracted with EtOAc. The extract was washed with H_2O, brine and dried over $MgSO_4$. The filtrate was concentrated under reduced pressure to give a pale yellow amorphous solid, which was used without further purification. To a stirred solution of this amorphous solid in Et_2O (1.5 mL) were added methyl propiolate (0.028 mL, 0.31 mmol) and Et_3N (0.043 mL, 0.31 mmol) at room temperature. The mixture was stirred overnight at room temperature. Concentration under pressure gave an oily residue, which was purified by flash chromatography over silica gel with *n*-hexane–EtOAc (4:1) to give **14** as a white amorphous solid (78.9 mg, 92% yield). IR (neat): 2122 (C≡C), 1709 (C=O), 1625 (C=C), 1373 (NSO_2), 1173 (NSO_2); ^1H NMR (400 MHz, $CDCl_3$) δ 2.35 (s, 3H), 2.59 (d, J = 2.1 Hz, 1H), 3.47 (dd, J = 13.9, 6.8 Hz, 1H), 3.53 (dd, J = 13.9, 6.8 Hz, 1H), 3.70 (s, 3H), 4.90 (ddd, J = 6.8, 6.8, 2.1 Hz, 1H), 5.38 (d, J = 12.4 Hz, 1H), 7.14 (dd, J = 8.0, 8.0 Hz, 1H), 7.22 (d, J = 8.3 Hz, 2H), 7.39 (d, J = 8.0 Hz, 1H), 7.55 (s, 1H), 7.57 (d, J = 12.4 Hz, 1H), 7.71 (d, J = 8.3 Hz, 2H), 7.97 (d, J = 8.0 Hz, 1H); ^{13}C NMR (125 MHz, $CDCl_3$) δ 21.5, 32.2, 51.2, 71.0, 76.8, 79.6, 99.1, 113.1, 114.0, 116.2, 125.6, 126.8 (2C), 127.5, 127.9, 128.2, 130.0 (2C), 134.7, 136.4, 145.4, 159.9, 167.7; HRMS (FAB) calcd $C_{23}H_{19}BrNO_5S$: $[M–H]^-$, 500.0173; found: 500.0174.

4.1.7 *(±)-(E)-4-Bromo-1-tosyl-3-{2-[3-(triisopropylsilyloxy) prop-1-enyloxy]but-3-ynyl}-1H-indole (15)*

To a stirred solution of the enol ether **14** (200 mg, 0.40 mmol) in Et_2O (6.5 mL) was added DIBAL-H (0.99 M solution in toluene; 1.0 mL, 1.0 mmol) at −78 °C. The mixture was stirred for 50 min at this temperature and quenched with 2 N Rochelle salt. After stirring for 1.5 h, the whole was extracted with Et_2O. The extract was washed with brine and dried over $MgSO_4$. The filtrate was concentrated under reduced pressure to give a crude alcohol as a white amorphous solid, which was used without further purification. To a stirred solution of this alcohol in DMF (2.0 mL) were added imidazole (81.7 mg, 1.2 mmol) and TIPSCl (0.127 mL,

0.60 mmol) at 0 °C. After stirring overnight at room temperature, the mixture was diluted with Et_2O. The organic phase was separated and washed with H_2O, brine and dried over $MgSO_4$. The filtrate was concentrated under reduced pressure to give an oily residue, which was purified by flash chromatography over silica gel with *n*-hexane–EtOAc (15:1) to give **15** as a colorless oil (239 mg, 95% yield). IR (neat): 2116 (C≡C), 1665 (C=C), 1369 (NSO_2), 1173 (NSO_2); ^{1}H NMR (500 MHz, $CDCl_3$) δ 1.04–1.09 (m, 21H), 2.34 (s, 3H), 2.48 (d, J = 2.3 Hz, 1H), 3.37 (dd, J = 13.7, 6.9 Hz, 1H), 3.52 (dd, J = 13.7, 6.9 Hz, 1H), 4.19 (d, J = 6.3 Hz, 2H), 4.75 (ddd, J = 6.9, 6.9, 2.3 Hz, 1H), 5.19 (dt, J = 12.0, 6.3 Hz, 1H), 6.48 (d, J = 12.0 Hz, 1H), 7.11 (dd, J = 8.3, 8.3 Hz, 1H), 7.21 (d, J = 8.0 Hz, 2H), 7.37 (d, J = 8.3 Hz, 1H), 7.57 (s, 1H), 7.72 (d, J = 8.0 Hz, 2H), 7.95 (d, J = 8.3 Hz, 1H); ^{13}C NMR (125 MHz, $CDCl_3$) δ 12.1 (3C), 18.0 (6C), 21.6, 32.4, 61.0, 69.2, 75.3, 81.2, 107.2, 113.0, 114.2, 117.1, 125.4, 126.9 (2C), 127.3, 127.9, 128.6, 129.9 (2C), 134.9, 136.3, 145.2, 145.5; HRMS (FAB) calcd $C_{31}H_{39}BrNO_4SSi$: $[M-H]^-$, 628.1558; found: 628.1555.

4.1.8 (±)-(2S,aR)-6-(4-Bromo-1-tosyl-1H-indol-3-yl) -2-(triisopropylsilyloxymethyl)hexa-3,4-dien-1-ol (16a) and (±)-(2R,aR)-Isomer (16b)

Microwave conditions (Table 4.1, entry 2): A solution of the silyl enol ether **15** (31 mg, 0.049 mmol) in $CHCl_3$ was heated under microwave irradiation at 120 °C for 12 min, then 150 °C for 12 min. The mixture was diluted with MeOH (0.4 mL), $NaBH_4$ (2.2 mg, 0.059 mmol) was added at room temperature. The mixture was stirred for 1 h at room temperature. Concentration under pressure gave an oily residue, which was purified by flash chromatography over silica gel with *n*-hexane–EtOAc (5:1) to give **16** as a colorless oil (25.4 mg, 82% yield, **a**:**b** = ca. 33:67).

Au-catalyzed conditions (Table 4.1, entry 3): To a stirred solution of the silyl enol ether **15** (50 mg, 0.079 mmol) in CH_2Cl_2 (0.25 mL) was added $[(Ph_3PAu)_3O]BF_4$ (4.3 mg, 0.004 mmol) at room temperature. After stirring for 7.5 h at 40 °C, the mixture was diluted with MeOH (0.5 mL). $NaBH_4$ (3.6 mg, 0.095 mmol) was added at room temperature, and the mixture was stirred for 1 h at room temperature. Concentration under pressure gave an oily residue, which was purified by flash chromatography over silica gel with *n*-hexane–EtOAc (5:1) to give **16** as a colorless oil (39.1 mg, 78% yield, **a**:**b** = ca. 80:20). Both diastereomers were isolated by HPLC [5C18-ARII column, 254 nm, MeCN:H_2O = 86:14, 8 mL/min; for analytical HPLC: 1 mL/min, t_1 = 48.25 min (minor isomer), t_2 = 49.80 min (major isomer)].

16a (major): IR (neat): 3456 (OH), 1963 (C=C=C), 1374 (NSO_2), 1173 (NSO_2); ^{1}H NMR (500 MHz, $CDCl_3$) δ 1.07–1.10 (m, 21H), 2.34 (s, 3H), 2.39–2.46 (m, 1H), 2.59 (dd, J = 6.3, 5.2 Hz, 1H), 3.52–3.72 (m, 5H), 3.78 (dd, J = 9.7,

4.6 Hz, 1H), 5.02 (ddd, $J = 9.7, 6.3, 2.9$ Hz, 1H), 5.45 (ddd, $J = 13.1, 6.3, 2.3$ Hz, 1H), 7.11 (dd, $J = 8.5, 8.0$ Hz, 1H), 7.23 (d, $J = 8.6$ Hz, 2H), 7.36 (dd, $J = 8.0, 1.1$ Hz, 1H), 7.44 (s, 1H), 7.73 (d, $J = 8.6$ Hz, 2H), 7.95 (dd, $J = 8.5, 1.1$ Hz, 1H); ^{13}C NMR (125 MHz, $CDCl_3$) δ 11.8 (3C), 17.9 (6C), 21.6, 26.5, 42.7, 65.9, 66.8, 89.8, 90.8, 112.9, 114.5, 121.8, 125.1, 125.4, 126.8 (2C), 127.7, 128.7, 129.9 (2C), 134.9, 136.5, 145.1, 204.7; HRMS (FAB) calcd $C_{31}H_{41}BrNO_4SSi$: $[M-H]^-$, 630.1714; found: 630.1707.

16b (minor): IR (neat): 3441 (OH), 1963 (C=C=C), 1375 (NSO_2), 1173 (NSO_2); ^{1}H NMR (500 MHz, $CDCl_3$) δ 1.02–1.07 (m, 21H), 2.35 (s, 3H), 2.38–2.46 (m, 1H), 2.60–2.67 (m, 1H), 3.55–3.72 (m, 5H), 3.78 (dd, $J = 9.7, 4.0$ Hz, 1H), 5.06 (ddd, $J = 9.7, 6.3, 2.9$ Hz, 1H), 5.44 (ddd, $J = 13.2, 6.3, 2.3$ Hz, 1H), 7.11 (dd, $J = 8.0, 8.0$ Hz, 1H), 7.23 (d, $J = 8.6$ Hz, 2H), 7.36 (d, $J = 8.0$ Hz, 1H), 7.43 (s, 1H), 7.73 (d, $J = 8.6$ Hz, 2H), 7.95 (d, $J = 8.0$ Hz, 1H); ^{13}C NMR (125 MHz, $CDCl_3$) δ 11.8 (3C), 17.9 (6C), 21.6, 26.5, 42.6, 66.0, 66.9, 89.8, 90.9, 112.9, 114.5, 121.8, 125.1, 125.4, 126.8 (2C), 127.7, 128.7, 129.9 (2C), 135.0, 136.5, 145.2, 204.7; HRMS (FAB) calcd $C_{31}H_{41}BrNO_4SSi$: $[M-H]^-$, 630.1714; found: 630.1705.

4.1.9 Determination of Relative Configuration of 16a: Synthesis of (±)-4-Bromo-1-tosyl-3-{[(2R,5S)-5-(triisopropylsilyloxymethyl)-5,6-dihydro-2H-pyran-2-yl]methyl}-1H-indole (17)

To a stirred suspension of $AgBF_4$ (3.1 mg, 0.016 mmol) in toluene (2.5 mL) was added Ph_3PAuCl (7.8 mg, 0.016 mmol) at room temperature. After stirring rapidly for 5 min, the resulting mixture was filtered through a cotton plug. To a solution of allenol **16a** (20 mg, 0.032 mmol) in toluene (0.25 mL) was added the above filtrate (0.25 mL) at room temperature. The resulting mixture was stirred for 8.5 h at this temperature. Concentration under pressure gave an oily residue, which was purified by flash chromatography over silica gel with *n*-hexane–EtOAc (15:1) to give **17** as a colorless oil (12.5 mg, 63% yield). IR (neat): 1598 (C=C), 1375 (NSO_2), 1173 (NSO_2); ^{1}H NMR (500 MHz, C_6D_6) δ 1.11–1.18 (m, 21H), 1.64 (s, 3H), 2.09–2.16 (br m, 1H), 3.16 (d, $J = 6.3$ Hz, 2H), 3.56 (dd, $J = 11.2, 3.7$ Hz, 1H), 3.72 (dd, $J = 9.2, 5.4$ Hz, 1H), 3.82 (dd, $J = 9.2, 9.2$ Hz, 1H), 4.15 (d, $J = 11.2$ Hz, 1H), 4.37–4.42 (m, 1H), 5.64 (d, $J = 10.6$ Hz, 1H), 5.70 (dd, $J = 10.6, 4.3$ Hz, 1H), 6.50 (d, $J = 8.6$ Hz, 2H), 6.71 (dd, $J = 8.0, 8.0$ Hz, 1H), 7.15 (d, $J = 8.0$ Hz, 1H), 7.61 (d, $J = 8.6$ Hz, 2H), 7.72 (s, 1H), 8.12 (d, $J = 8.0$ Hz, 1H); ^{13}C NMR (125 MHz, $CDCl_3$) δ 12.0 (3C), 18.1 (6C), 21.6, 32.1, 38.3, 64.4, 73.5, 77.2, 113.0, 114.5, 119.4, 125.2, 126.4, 126.6, 126.9 (2C), 127.9, 129.0, 129.9 (2C), 131.0, 135.0, 136.5, 145.1; HRMS (FAB) calcd $C_{31}H_{41}BrNO_4SSi$: $[M+H]^+$, 630.1714; found: 630.1711.

4.1.10 *N-[(2S,aR)-6-(4-Bromo-1-tosyl-1H-indol-3-yl)-2-(triisopropylsilyloxymethyl)hexa-3,4-dienyl]-2-nitrobenzenesulfonamide (18a) and Its (±)-(2R,aR)-Isomer (18b)*

To a stirred mixture of the allenol **16** (**a**:**b** = ca. 80:20) (300 mg, 0.48 mmol), $NsNH_2$ (317 mg, 1.57 mmol) and PPh_3 (630 mg, 2.40 mmol) in benzene (18 mL) was added diethyl azodicarboxylate (40% solution in toluene; 1.10 mL, 2.40 mmol) at room temperature, and the mixture was stirred for 1.5 h at this temperature. Concentration under pressure gave an oily residue, which was purified by flash chromatography over silica gel with *n*-hexane–EtOAc (3:1) to give **18** as a pale yellow amorphous solid (276 mg, 70% yield, **a**:**b** = 80:20).

18a (major): IR (neat): 1962 (C=C=C), 1540 (NO_2), 1372 (NSO_2), 1172 (NSO_2); ^{1}H NMR (500 MHz, $CDCl_3$) δ 0.99–1.07 (m, 21H), 2.35 (s, 3H), 2.36–2.42 (m, 1H), 3.05 (ddd, J = 12.6, 6.3, 5.1 Hz, 1H), 3.27 (ddd, J = 12.6, 6.3, 5.3 Hz, 1H), 3.39 (dd, J = 10.3, 8.0 Hz, 1H), 3.56–3.70 (m, 2H), 3.61 (dd, J = 10.3, 4.6 Hz, 1H), 5.04 (ddd, J = 9.7, 6.3, 2.9 Hz, 1H), 5.49 (ddd, J = 13.2, 6.3, 2.3 Hz, 1H), 5.67 (t, J = 6.3 Hz, 1H), 7.10 (dd, J = 8.0, 8.0 Hz, 1H), 7.23 (d, J = 8.6 Hz, 2H), 7.34 (d, J = 8.0 Hz, 1H), 7.38 (s, 1H), 7.64-7.70 (m, 2H), 7.72 (d, J = 8.6 Hz, 2H), 7.79 (dd, J = 7.4, 1.7 Hz, 1H), 7.94 (d, J = 8.0 Hz, 1H), 8.09 (dd, J = 7.2, 2.0 Hz, 1H); ^{13}C NMR (125 MHz, $CDCl_3$) δ 11.8 (3C), 18.0 (6C), 21.6, 26.5, 41.5, 45.2, 65.1, 90.1, 91.8, 112.9, 114.4, 121.7, 125.0, 125.2, 125.5, 126.8 (2C), 127.7, 128.6, 130.0 (2C), 131.0, 132.6, 133.4, 133.8, 134.9, 136.5, 145.3, 148.0, 204.5; HRMS (FAB) calcd $C_{37}H_{45}BrN_3O_7S_2Si$: $[M–H]^-$, 814.1657; found: 814.1662.

18b (minor): IR (neat): 1963 (C=C=C), 1541 (NO_2), 1372 (NSO_2), 1172 (NSO_2); ^{1}H NMR (500 MHz, $CDCl_3$) δ 0.99–1.05 (m, 21H), 2.31–2.34 (m, 1H), 2.35 (s, 3H), 3.10 (ddd, J = 12.6, 6.3, 5.7 Hz, 1H), 3.30 (ddd, J = 12.6, 6.3, 6.3 Hz, 1H), 3.40 (dd, J = 9.7, 7.4 Hz, 1H), 3.61 (dd, J = 9.7, 4.9 Hz, 1H), 3.61–3.64 (m, 2H), 5.00 (ddd, J = 9.7, 6.9, 3.4 Hz, 1H), 5.41 (ddd, J = 13.2, 6.9, 1.7 Hz, 1H), 5.65 (t, J = 6.3 Hz, 1H), 7.10 (dd, J = 8.0, 8.0 Hz, 1H), 7.24 (d, J = 8.0 Hz, 2H), 7.34 (d, J = 8.0 Hz, 1H), 7.38 (s, 1H), 7.61–7.69 (m, 2H), 7.73 (d, J = 8.0 Hz, 2H), 7.77 (dd, J = 7.6, 1.1 Hz, 1H), 7.94 (d, J = 8.0 Hz, 1H), 8.09 (dd, J = 7.4, 1.7 Hz, 1H); ^{13}C NMR (125 MHz, $CDCl_3$) δ 11.8 (3C), 18.0 (6C), 21.6, 26.4, 41.6, 45.5, 65.1, 90.0, 91.5, 112.9, 114.5, 121.6, 125.1, 125.2, 125.4, 126.8 (2C), 127.7, 128.7, 130.0 (2C), 131.0, 132.6, 133.3, 133.8, 134.9, 136.5, 145.3, 148.0, 204.8; HRMS (FAB) calcd $C_{37}H_{45}BrN_3O_7S_2Si$: $[M–H]^-$, 814.1657; found: 814.1655.

4.1.11 (±)-N-[(2S,aR)-6-(4-Bromo-1-tosyl-1H-indol-3-yl)-2-(triisopropylsilyloxymethyl)-hexa-3,4-dienyl]-4-methylbenzenesulfonamide (19a) and Its (±)-(2R,aR)-Isomer (19b)

To a stirred mixture of the allenol **16** (**a**:**b** = ca. 80:20; 150 mg, 0.24 mmol), FmocNHTs (308 mg, 0.78 mmol) and PPh_3 (312 mg, 1.19 mmol) in THF (4 mL) was added diethyl azodicarboxylate (0.54 mL, 1.19 mmol; 40% solution in toluene) at 0 °C, and the mixture was stirred for 3 h at room temperature. Concentration under pressure gave an oily residue, which was dissolved in DMF (7 mL). Piperidine (94 μL, 0.95 mmol) was added to the mixture at 0 °C. After stirring for 50 min at room temperature, the mixture was diluted with Et_2O and washed with H_2O, brine and dried over $MgSO_4$. The filtrate was concentrated under reduced pressure to give an oily residue, which was purified by flash chromatography over silica gel with *n*-hexane–EtOAc (2:1) to give **19** as a yellow amorphous solid (136 mg, 73% yield, **a**:**b** = ca. 80:20).

19a (major): IR (neat): 1964 (C=C=C), 1374 (NSO_2), 1173 (NSO_2); ^{1}H NMR (500 MHz, $CDCl_3$) δ 0.98–1.04 (m, 21H), 2.29–2.34 (m, 1H), 2.34 (s, 3H), 2.40 (s, 3H), 2.98 (dd, J = 6.0, 6.0 Hz, 1H), 3.00 (dd, J = 6.0, 6.0 Hz, 1H), 3.36 (dd, J = 10.0, 8.3 Hz, 1H), 3.60 (dd, J = 10.0, 4.3 Hz, 1H), 3.60–3.64 (m, 2H), 4.96 (ddd, J = 9.1, 6.3, 2.9 Hz, 1H), 5.14 (t, J = 6.0 Hz, 1H), 5.44 (ddd, J = 13.2, 6.3, 2.3 Hz, 1H), 7.10 (dd, J = 8.0, 8.0 Hz, 1H), 7.23 (d, J = 8.6 Hz, 2H), 7.28 (d, J = 8.6 Hz, 2H), 7.34 (d, J = 8.0 Hz, 1H), 7.41 (s, 1H), 7.70–7.74 (m, 4H), 7.94 (d, J = 8.0 Hz, 1H); ^{13}C NMR (125 MHz, $CDCl_3$) δ 11.7 (3C), 17.9 (6C), 21.5, 21.6, 26.4, 40.5, 45.9, 66.4, 90.2, 91.6, 112.9, 114.4, 121.6, 125.1, 125.5, 126.8 (2C), 127.1 (2C), 127.7, 128.6, 129.6 (2C), 130.0 (2C), 134.9, 136.5, 137.0, 143.2, 145.2, 204.5; HRMS (FAB) calcd $C_{38}H_{48}BrN_2O_5S_2Si$: $[M–H]^-$, 783.1963; found: 783.1960.

19b (minor): IR (neat): 1964 (C=C=C), 1374 (NSO_2), 1173 (NSO_2); ^{1}H NMR (500 MHz, $CDCl_3$) δ 0.94–1.04 (m, 21H), 2.27–2.33 (m, 1H), 2.35 (s, 3H), 2.41 (s, 3H), 2.99 (ddd, J = 6.3, 6.3, 1.7 Hz, 1H), 3.01 (ddd, J = 6.3, 6.3, 1.7 Hz, 1H), 3.35 (dd, J = 10.0, 7.7 Hz, 1H), 3.60 (dd, J = 10.0, 4.3 Hz, 1H), 3.61–3.65 (m, 2H), 4.94 (ddd, J = 9.7, 6.3, 2.9 Hz, 1H), 5.11 (t, J = 6.3 Hz, 1H), 5.42 (ddd, J = 13.2, 6.3, 2.3 Hz, 1H), 7.10 (dd, J = 8.2, 8.2 Hz, 1H), 7.23 (d, J = 8.0 Hz, 2H), 7.28 (d, J = 8.0 Hz, 2H), 7.35 (d, J = 8.2 Hz, 1H), 7.41 (s, 1H), 7.71 (d, J = 8.0 Hz, 2H), 7.73 (d, J = 8.0 Hz, 2H), 7.94 (d, J = 8.2 Hz, 1H); ^{13}C NMR (125 MHz, $CDCl_3$) δ 11.7 (3C), 17.9 (6C), 21.5, 21.6, 26.3, 40.5, 46.0, 66.4, 90.2, 91.5, 112.9, 114.4, 121.5, 125.1, 125.4, 126.8 (2C), 127.1 (2C), 127.7, 128.6, 129.6 (2C), 130.0 (2C), 134.9, 136.5, 137.0, 143.2, 145.2, 204.6; HRMS (FAB) calcd $C_{38}H_{48}BrN_2O_5S_2Si$: $[M–H]^-$, 783.1963; found: 783.1968.

4.1.12 (±)-(6aR,9S)-7-(2-Nitrophenylsulfonyl)-4-tosyl-9-(triisopropylsilyloxymethyl)-4,6,6a,7,8,9-hexahydroindolo[4,3-fg]quinoline (20a) and Its (±)-(6aS,9S)-Isomer (20b) (Table 2, Entry 3)

To a stirred mixture of allenamide **18** (**a**:**b** = 80:20; 30 mg, 0.037 mmol) in DMF (0.6 mL) were added $Pd(PPh_3)_4$ (2.1 mg, 0.0018 mmol) and K_2CO_3 (15 mg, 0.11 mmol) at room temperature under argon, and the mixture was stirred for 3.5 h at 100 °C. The mixture was diluted with Et_2O and washed with H_2O, brine and dried over $MgSO_4$. The filtrate was concentrated under reduced pressure to give a yellow oil which was purified by flash chromatography over silica gel with *n*-hexane–EtOAc (5:1) to give **20** as a yellow amorphous solid (22.3 mg, 83% yield, **a**:**b** = 73:27). Both diastereomers were isolated by PTLC with hexane–$(i\text{-Pr})_2O$ (3:1).

20a (major): IR (neat): 1596 (C=C), 1544 (NO_2), 1359 (NSO_2), 1178 (NSO_2); ^1H NMR (500 MHz, $CDCl_3$) δ 0.97–1.04 (m, 21H), 2.36 (s, 3H), 2.40–2.48 (br m, 1H), 2.95 (dd, J = 13.7, 10.9 Hz, 1H), 2.99 (ddd, J = 14.9, 12.0, 2.3 Hz, 1H), 3.27 (dd, J = 14.9, 5.2 Hz, 1H), 3.55 (dd, J = 9.7, 8.0 Hz, 1H), 3.69 (dd, J = 9.7, 5.7 Hz, 1H), 4.10 (dd, J = 13.7, 5.2 Hz, 1H), 4.75–4.80 (m, 1H), 6.16 (s, 1H), 7.18–7.21 (m, 2H), 7.23–7.30 (m, 3H), 7.60–7.65 (m, 2H), 7.66–7.71 (m, 1H), 7.78 (d, J = 8.0 Hz, 2H), 7.79 (d, J = 7.4 Hz, 1H), 8.04 (dd, J = 8.0, 1.1 Hz, 1H); ^{13}C NMR (125 MHz, $CDCl_3$) δ 11.8 (3C), 17.9 (6C), 21.6, 29.7, 38.2, 43.0, 54.1, 64.9, 112.7, 115.6, 117.3, 120.5, 124.0, 124.3, 125.8, 126.8 (2C), 128.2, 130.0 (2C), 130.2, 131.0, 131.8, 133.4, 133.5, 133.6, 133.8, 135.4, 144.9, 147.9; HRMS (FAB) calcd $C_{37}H_{44}N_3O_7S_2Si$: $[M-H]^-$, 734.2395; found: 734.2392.

20b (minor): IR (neat): 1597 (C=C), 1542 (NO_2), 1359 (NSO_2), 1174 (NSO_2); ^1H NMR (500 MHz, $CDCl_3$) δ 0.94–1.01 (m, 21H), 2.36 (s, 3H), 2.53–2.58 (br m, 1H), 2.99 (ddd, J = 14.3, 12.0, 2.3 Hz, 1H), 3.12 (dd, J = 14.3, 5.0 Hz, 1H), 3.33 (dd, J = 9.7, 8.0 Hz, 1H), 3.35 (dd, J = 13.7, 3.4 Hz, 1H), 3.48 (dd, J = 9.7, 6.9 Hz, 1H), 3.97 (d, J = 13.7 Hz, 1H), 4.74 (dd, J = 12.0, 5.0 Hz, 1H), 6.31 (d, J = 5.2 Hz, 1H), 7.15 (d, J = 2.3 Hz, 1H), 7.18 (d, J = 6.9 Hz, 1H), 7.24 (d, J = 8.6 Hz, 2H), 7.28 (dd, J = 8.0, 8.0 Hz, 1H), 7.62–7.72 (m, 3H), 7.76 (d, J = 8.6 Hz, 2H), 7.78 (d, J = 8.0 Hz, 1H), 8.14 (dd, J = 7.4, 1.7 Hz, 1H); ^{13}C NMR (125 MHz, $CDCl_3$) δ 11.9 (3C), 18.0 (6C), 21.6, 28.7, 39.5, 40.3, 53.9, 63.9, 112.7, 115.7, 117.2, 120.4, 123.1, 124.3, 125.8, 126.8 (2C), 128.2, 130.0 (2C), 130.6, 131.4, 131.9, 133.3, 133.6, 134.0, 134.1, 135.4, 144.9, 147.8; HRMS (FAB) calcd $C_{37}H_{44}N_3O_7S_2Si$: $[M-H]^-$, 734.2395; found: 734.2392.

4.1.13 (±)-(6aR,9S)-4,7-Ditosyl-9-(triisopropylsilyloxymethyl)-4,6,6a,7,8,9-hexahydroindo-lo[4,3-fg]quinoline (21a) and Its (±)-(6aS,9S)-Isomer (21b) (Table 4.2, Entry 12)

To a stirred mixture of allenamide **19** (**a**:**b** = ca. 80:20; 30 mg, 0.038 mmol) in DMF (0.6 mL) were added $Pd(PPh_3)_4$ (2.2 mg, 0.0019 mmol) and K_2CO_3 (15.8 mg, 0.11 mmol) at room temperature under argon, and the mixture was stirred for 3 h at 120 °C. The mixture was diluted with Et_2O and washed with H_2O, brine and dried over $MgSO_4$. The filtrate was concentrated under reduced pressure to give a yellow oil which was purified by flash chromatography over silica gel with *n*-hexane–EtOAc (6:1) to give **21** as a white amorphous solid (17.3 mg, 65% yield, **a**:**b** = 87:13). Both diastereomers were isolated by PTLC with hexane–(i-Pr)$_2$O (1:1).

21a (major): IR (neat): 1598 (C=C), 1376 (NSO_2), 1178 (NSO_2); ^1H NMR (500 MHz, $CDCl_3$) δ 0.95–1.05 (m, 21H), 2.12–2.19 (br m, 1H), 2.35 (s, 3H), 2.39 (s, 3H), 2.84 (dd, J = 13.7, 10.6 Hz, 1H), 2.92 (ddd, J = 14.0, 12.0, 1.7 Hz, 1H), 3.33 (dd, J = 14.0, 5.4 Hz, 1H), 3.46 (dd, J = 9.6, 8.6 Hz, 1H), 3.63 (dd, J = 9.6, 5.4 Hz, 1H), 4.11 (dd, J = 13.7, 5.2 Hz, 1H), 4.67–4.73 (m, 1H), 6.07 (s, 1H), 7.15 (d, J = 7.4 Hz, 1H), 7.20–7.28 (m, 6H), 7.67 (d, J = 8.0 Hz, 2H), 7.77 (d, J = 8.0 Hz, 2H), 7.78 (d, J = 7.4 Hz, 1H); ^{13}C NMR (125 MHz, $CDCl_3$) δ 11.9 (3C), 17.9 (6C), 21.5, 21.6, 30.0, 37.3, 42.9, 53.6, 65.0, 112.7, 115.5, 117.7, 120.5, 124.0, 125.7, 126.8 (2C), 126.9 (2C), 128.3, 129.8 (2C), 129.9 (2C), 130.2, 133.3, 133.4, 135.5, 138.0, 143.3, 144.8; HRMS (FAB) calcd $C_{38}H_{49}N_2O_5S_2Si$: $[M+H]^+$, 705.2852; found: 705.2850.

21b (minor): IR (neat): 1598 (C=C), 1377 (NSO_2), 1173 (NSO_2), ^1H NMR (500 MHz, $CDCl_3$) δ 0.99–1.05 (m, 21H), 2.35 (s, 3H), 2.40 (s, 3H), 2.49–2.55 (br m, 1H), 2.87 (ddd, J = 14.3, 12.0, 1.7 Hz, 1H), 3.21 (dd, J = 13.2, 3.7 Hz, 1H), 3.30–3.39 (m, 3H), 3.89 (d, J = 13.2 Hz, 1H), 4.64 (dd, J = 11.5, 4.6 Hz, 1H), 6.29 (d, J = 5.2 Hz, 1H), 7.13 (d, J = 7.4 Hz, 1H), 7.18 (s, 1H), 7.22–7.29 (m, 5H), 7.72 (d, J = 8.0 Hz, 2H), 7.77 (d, J = 8.0 Hz, 2H), 7.78 (d, J = 8.0 Hz, 1H); ^{13}C NMR (125 MHz, $CDCl_3$) δ 11.9 (3C), 18.0 (6C), 21.5, 21.6, 28.4, 39.5, 39.9, 53.7, 64.3, 112.7, 115.6, 117.6, 120.5, 123.6, 125.8, 126.8 (2C), 127.1 (2C), 128.3, 129.7 (2C), 129.9 (2C), 130.7, 133.3, 134.1, 135.5, 138.1, 143.2, 144.8; HRMS (FAB) calcd $C_{38}H_{49}N_2O_5S_2Si$: $[M+H]^+$, 705.2852; found: 705.2849.

Determination of relative configuration of 21a: To a stirred mixture of **20a** (25 mg, 0.034 mmol) in DMF (0.2 mL) were added $LiOH{\cdot}H_2O$ (14.3 mg 0.34 mmol) and $HSCH_2CO_2H$ (11.8 μL, 0.17 mmol) at 0 °C. After stirring for 1 h at room temperature, the mixture was diluted with EtOAc was washed with saturated $NaHCO_3$, brine and dried over $MgSO_4$. The filtrate was concentrated under reduced pressure to give a crude amine as an oily residue, which was used without further purification. To a stirred solution of this amine in CH_2Cl_2 (0.25 mL) were added Et_3N (14.2 μL, 0.102 mmol) and TsCl (9.7 mg, 0.051 mmol) at 0 °C. After stirring for 2 h at room temperature, the mixture was diluted with EtOAc and washed with H_2O, brine and dried over $MgSO_4$. The filtrate was concentrated under reduced pressure to give an oily residue, which was purified by flash chromatography over silica gel with *n*-hexane–EtOAc (6:1) to give **21a** as a white amorphous solid (18.1 mg, 76% yield).

4.1.14 (±)-(6aR,9S)-7-Methyl-4-tosyl-9-(triisopropylsilyloxymethyl)-4,6,6a,7,8,9-hexahydroindolo[4,3-fg]quinoline (22a) and Its (±)-(6aS,9S)-Isomer (22b)

To a stirred mixture of **20** (**a**:**b** = 74:26) (136 mg, 0.19 mmol) in DMF (1.1 mL) were added LiOH·H_2O (78 mg 1.9 mmol) and $HSCH_2CO_2H$ (64 μL, 0.92 mmol) at 0 °C. After stirring for 1 h at room temperature, the mixture was diluted with EtOAc and washed with saturated $NaHCO_3$, brine and dried over $MgSO_4$. The filtrate was concentrated under reduced pressure to give a crude amine as an oily residue, which was used without further purification. To a stirred solution of this amine in DMF (2.0 mL) were added K_2CO_3 (41 mg, 0.30 mmol) and MeI (15 μL, 0.24 mmol) at 0 °C. After stirring for 5 h at room temperature, the mixture was diluted with EtOAc and washed with saturated $NaHCO_3$, brine and dried over $MgSO_4$. The filtrate was concentrated under reduced pressure to give an oily residue, which was purified by flash chromatography over silica gel with *n*-hexane–EtOAc (5:1 to 3:1) to give **22a** (53.4 mg, 52% yield) and **22b** (16.7 mg, 16% yield) both as a brown amorphous solid.

22a: IR (neat): 1599 (C=C), 1379 (NSO_2), 1177 (NSO_2), ^1H NMR (500 MHz, $CDCl_3$) δ 1.03–1.09 (m, 21H), 2.33 (s, 3H), 2.46 (s, 3H), 2.48–2.53 (m, 3H), 2.95–3.04 (m, 2H), 3.37 (dd, J = 15.4, 5.4 Hz, 1H), 3.72 (dd, J = 9.3, 5.2 Hz, 1H), 3.78 (dd, J = 9.3, 9.0 Hz, 1H), 6.37 (d, J = 3.9 Hz, 1H), 7.16–7.21 (m, 1H), 7.19 (d, J = 8.3 Hz, 2H), 7.23–7.29 (m, 2H), 7.72–7.76 (m, 1H), 7.74 (d, J = 8.3 Hz, 2H); ^{13}C NMR (125 MHz, $CDCl_3$) δ 12.0 (3C), 18.0 (6C), 21.5, 27.2, 39.2, 43.7, 53.0, 62.2, 65.0, 112.2, 116.1, 118.4, 119.7, 123.2, 125.8, 126.7 (2C), 128.6, 129.8 (2C), 129.9, 133.5, 135.0, 135.5, 144.6; HRMS (FAB) calcd $C_{32}H_{43}N_2O_3SSi$: $[M-H]^-$, 563.2769; found: 563.2770.

22b: IR (neat): 1599 (C=C), 1379 (NSO_2), 1178 (NSO_2), ^1H NMR (500 MHz, $CDCl_3$) δ 1.05–1.09 (m, 21H), 2.22 (dd, J = 10.7, 10.7 Hz, 1H), 2.33 (s, 3H), 2.50–2.58 (m, 4H), 2.82–2.93 (m, 1H), 2.98–3.01 (m, 1H), 3.07 (dd, J = 11.1, 5.0 Hz, 1H), 3.43 (dd, J = 15.1, 5.4 Hz, 1H), 3.65 (dd, J = 9.5, 7.6 Hz, 1H), 3.71 (dd, J = 9.5, 6.3 Hz, 1H), 6.38 (s, 1H), 7.19 (s, 1H), 7.20 (d, J = 8.3 Hz, 2H), 7.24–7.30 (m, 2H), 7.73–7.77 (m, 1H), 7.76 (d, J = 8.3 Hz, 2H); ^{13}C NMR (125 MHz, $CDCl_3$) δ 12.0 (3C), 18.0 (6C), 21.5, 27.0, 39.4, 44.0, 56.8, 62.5, 65.7, 112.2, 116.3, 118.1, 119.7, 123.9, 125.8, 126.7 (2C), 128.5, 129.6, 129.8, 133.5 (2C), 133.8, 135.6, 144.6; HRMS (FAB) calcd $C_{32}H_{43}N_2O_3SSi$: $[M-H]^-$, 563.2769; found: 563.2770.

(±)-Isolysergol (3). To a stirred solution of **22a** (8.3 mg, 0.015 mmol) in THF (0.33 mL) was added TBAF (1.00 M solution in THF; 18 μL, 0.018 mmol) at 0 °C. The mixture was stirred for 1 h at room temperature and quenched with H_2O. The whole was extracted with EtOAc. The extract was washed with H_2O, brine and dried over $MgSO_4$. The filtrate was concentrated under reduced pressure to give a crude alcohol as a brown amorphous solid, which was used without further purification. To a stirred solution of this alcohol in MeOH (0.45 mL) was added Mg (3.6 mg, 0.15 mmol) at room temperature. The mixture was stirred for 2 h at this temperature. Concentration under pressure gave an oily residue, which was purified by PTLC with EtOAc–MeOH (3:1) to give isolysergol (**3**) as a pale brown solid (3.8 mg, 99% yield). IR (neat): 3213 (OH), 1604 (C=C), The IR spectra was found to be identical with that of natural isolysergol [4]. ^{1}H NMR (500 MHz, $CDCl_3$–CD_3OD) δ 2.44-2.50 (m, 1H), 2.55 (s, 3H), 2.65 (ddd, J = 14.3, 11.5, 1.7 Hz, 1H), 2.85 (ddd, J = 11.5, 4.0, 1.7 Hz, 1H), 3.04 (d, J = 11.5 Hz, 1H), 3.14-3.19 (m, 1H), 3.53 (dd, J = 14.3, 5.7 Hz, 1H), 3.80 (ddd, J = 10.3, 3.6, 1.7 Hz, 1H), 3.96 (dd, J = 10.3, 3.4 Hz, 1H), 6.46 (d, J = 5.7 Hz, 1H), 6.89 (d, J = 1.7 Hz, 1H), 7.14-7.17 (m, 2H), 7.18-7.22 (m, 1H); The ^{1}H NMR spectra was found to be identical with that of synthesized isolysergol reported by Ninomiya et al. [20]. ^{13}C NMR (125 MHz, $CDCl_3$–CD_3OD) δ 27.3, 36.3, 43.3, 57.4, 63.0, 66.0, 109.5, 109.9, 111.7, 118.2, 121.0, 122.9, 126.0, 128.0, 133.8, 136.7; HRMS (FAB) calcd $C_{16}H_{17}N_2O$: $[M–H]^-$, 253.1346; found: 253.1352.

(±)-Lysergol (2). To a stirred solution of **22b** (16.7 mg, 0.030 mmol) in THF (0.7 mL) was added TBAF (1.00 M solution in THF; 39 μL, 0.039 mmol) at 0 °C. The mixture was stirred for 1.5 h at room temperature and quenched with H_2O. The whole was extracted with EtOAc. The extract was washed with H_2O, brine and dried over $MgSO_4$. The filtrate was concentrated under reduced pressure to give a crude alcohol as a brown amorphous solid, which was used without further purification. To a stirred solution of this alcohol in MeOH (0.85 mL) was added Mg (7.3 mg, 0.30 mmol) at room temperature. The mixture was stirred for 3 h at this temperature. Concentration under pressure gave an oily residue, which was purified by PTLC with EtOAc–MeOH (2:1) to give lysergol (**2**) as a pale brown solid (7.0 mg, 92% yield). IR (neat): 3427 (OH), 1606 (C=C), ^{1}H NMR (500 MHz, $CDCl_3$–CD_3OD) δ 2.36 (dd, J = 10.9, 10.9 Hz, 1H), 2.61 (s, 3H), 2.74 (ddd, J = 13.7, 12.0, 1.7 Hz, 1H), 2.85–2.93 (m, 1H), 3.17 (dd, J = 10.9, 5.2 Hz, 1H), 3.23–3.30 (m, 1H), 3.51–3.59 (m, 2H), 3.70 (dd, J = 10.9, 5.7 Hz, 1H), 6.41 (s, 1H), 6.94 (s, 1H), 7.13–7.18 (m, 2H), 7.20–7.25 (m, 1H); ^{13}C NMR (125 MHz, $CDCl_3$–CD_3OD) δ 26.3, 38.1, 43.2, 56.5, 63.1, 64.6, 109.5 (2C), 111.6, 118.4, 121.0, 122.8, 125.8, 127.6, 133.9, 135.0; The IR, ^{1}H NMR and ^{13}C NMR spectra were found to be identical with those of natural lysergol. HRMS (FAB) calcd $C_{16}H_{17}N_2O$: $[M–H]^-$, 253.1346; found: 253.1349.

4.1.15 Methyl (±)-(6aR,9S)-4,7-ditosyl-4,6,6a,7,8,9-hexahydroindolo[4,3-fg]quinoline-9-carboxylate (23a)

To a stirred solution of **21** (**a**:**b** = 87:13) (190 mg, 0.27 mmol) in THF (5 mL) was added TBAF (1.00 M solution in THF; 0.32 mL, 0.32 mmol) at 0 °C. The mixture was stirred for 40 min at room temperature and quenched with H_2O. The whole was extracted with EtOAc. The extract was washed with H_2O, brine and dried over $MgSO_4$. Concentration of the filtrate under reduced pressure followed by filtration through a short pad of SiO_2 with EtOAc give a crude alcohol. To a stirred solution of this alcohol in CH_2Cl_2 (10 mL) was added Dess-Martin periodinane (230 mg, 0.54 mmol) at 0 °C. After stirring for 30 min at this temperature, the mixture was warming to room temperature. The mixture was stirred for further 1 h at this temperature and filtrated through a short pad of SiO_2 with EtOAc to give a crude aldehyde. To a stirred mixture of the crude aldehyde and 2-methylbut-2-ene (1.66 mL, 16.2 mmol) in a mixed solvent of THF (2.9 mL) and *t*-BuOH (2.9 mL) were added $NaClO_2$ (117 mg, 1.30 mmol) and NaH_2PO_4 (155 mg, 1.30 mmol) at room temperature. After stirring for 1.5 h at room temperature, brine was added to the mixture. The whole was extracted with EtOAc. The extract was washed with brine and dried over $MgSO_4$. The filtrate was concentrated under reduced pressure to give a crude carboxylic acid. To a stirred solution of this acid in a mixed solvent of toluene (1.7 mL) and MeOH (1.2 mL) was added $TMSCHN_2$ (2.00 M solution in Et_2O; 0.35 mL, 0.70 mmol) at 0 °C. The mixture was stirred for 30 min at room temperature. Concentration under pressure gave an oily residue, which was purified by flash chromatography over silica gel with *n*-hexane–EtOAc (4:1) to give **23a** as a pale yellow amorphous solid (96.4 mg, 62% yield). IR (neat): 1736 (C=O), 1597 (C=C), 1347 (NSO_2), 1177 (NSO_2), ^1H NMR (500 MHz, $CDCl_3$) δ 2.35 (s, 3H), 2.42 (s, 3H), 2.92 (ddd, J = 14.9, 12.0, 2.3 Hz, 1H), 3.03–3.08 (m, 1H), 3.19 (dd, J = 14.3, 10.9 Hz, 1H), 3.27 (dd, J = 14.9, 5.2 Hz, 1H), 3.70 (s, 3H), 4.26 (dd, J = 14.3, 5.2 Hz, 1H), 4.69–4.75 (m, 1H), 6.37 (s, 1H), 7.18–7.30 (m, 7H), 7.69 (d, J = 8.0 Hz, 2H), 7.76 (d, J = 8.0 Hz, 2H), 7.81 (d, J = 8.6 Hz, 1H); ^{13}C NMR (125 MHz, $CDCl_3$) δ 21.5, 21.6, 29.4, 40.2, 40.8, 52.3, 53.0, 113.1, 115.9, 117.2, 120.4, 120.7, 125.8, 126.7 (4C), 128.3, 129.6, 129.9 (2C), 130.0 (2C), 133.4, 134.1, 135.4, 137.8, 143.7, 144.9, 171.2; HRMS (FAB) calcd for $C_{30}H_{29}N_2O_6S_2$: $[M+H]^+$, 577.1467; found: 577.1471.

Determination of relative configuration of 23a: To a stirred solution of **23a** (5.0 mg, 0.0086 mmol) in MeOH (0.5 mL) was added $NaBH_4$ (1.63 mg, 0.043 mmol) at room temperature [31]. After stirring for 1 h at this temperature, H_2O was added, and the mixture was concentrated under reduced pressure. The residue was dissolved in EtOAc and washed with brine and dried over $MgSO_4$. The filtrate was concentrated under reduced pressure to give a crude alcohol, which was used without further purification. To a stirred solution of this alcohol in DMF (0.2 mL) were added imidazole (16.6 mg, 0.24 mmol) and TIPSCl

(0.026 mL, 0.12 mmol) at 0 °C. After stirring overnight at room temperature, the mixture was diluted with Et_2O and washed with H_2O, brine and dried over $MgSO_4$. The filtrate was concentrated under reduced pressure to give an oily residue, which was purified by PTLC with *n*-hexane–EtOAc (3:1) to give **21a** as a white amorphous solid (4.1 mg, 68% yield).

4.1.16 (±)-Methyl Isolysergate (24a) and (±)-Methyl Lysergate (24b)

To a stirred solution of **23a** (30 mg, 0.052 mmol) in THF (1.6 mL) was added sodium naphthalenide (0.67 M solution in THF; 0.78 mL, 0.52 mmol) [33] at – 78 °C under argon. The mixture was stirred for 10 min at this temperature and quenched with saturated NH_4Cl. The mixture was made basic with saturated $NaHCO_3$. The whole was extracted with EtOAc. The extract was washed with brine and dried over $MgSO_4$. Concentration of the filtrate under reduced pressure gave a crude amine which was used without further purification. To a stirred solution of this amine in MeOH (3.0 mL) were added formalin (0.02 mL, 0.26 mmol), $NaBH_3CN$ (16.3 mg, 0.26 mmol) and AcOH (55 μL) at room temperature. After stirring for 1.5 h at this temperature, the mixture was quenched with saturated $NaHCO_3$. The mixture was concentrated under pressure followed by filtration through a short pad of SiO_2 with EtOAc. The filtrate was concentrated under reduced pressure to give a yellow solid, which was purified by flash chromatography over silica gel with *n*-hexane–EtOAc (1:10) to give **24a** and **24b** as a yellow solid (9.0 mg, 61% yield, **a**:**b** = 35:65). ^1H NMR spectra of **24a** and **24b** were in agreement with those reported by Hendrickson and Wang [15]: ^1H NMR (400 MHz, $CDCl_3$) of methyl lysergate **24b** (major isomer): δ 2.63 (s, 3H), 2.68–2.73 (m, 2H), 3.20–3.27 (m, 1H), 3.30 (dd, J = 11.6, 4.9 Hz, 1H), 3.53 (dd, J = 14.5, 5.5 Hz, 1H), 3.73–3.76 (m, 1H), 3.79 (s, 3H), 6.60 (s, 1H), 6.91 (t, J = 1.8 Hz, 1H), 7.16–7.25 (m, 3H), 7.92 (br s, 1H); methyl isolysergate **24a** (minor isomer): δ 2.59 (s, 3H), 2.75–2.81 (m, 2H), 3.20–3.27 (m, 1H), 3.29–3.34 (m, 1H), 3.38 (dd, J = 11.6, 3.0 Hz, 1H), 3.44 (dd, J = 14.6, 5.4 Hz, 1H), 3.73 (s, 3H), 6.56 (d, J = 5.4 Hz, 1H), 6.91 (t, J = 1.8 Hz, 1H), 7.16–7.25 (m, 3H), 7.92 (br s, 1H); IR (neat): 3410 (NH), 1731 (C=O), 1604 (C=C); HRMS (FAB) calcd $C_{17}H_{17}N_2O_2$: $[M–H]^-$, 281.1296; found: 281.1304.

(±)-Lysergic Acid (1). The preparation of lysergic acid (**1**) was carried out according to the method of Hendrickson and Wang [15] and Szántay [16]: To solution of diastereomixture of methyl lysergate and isolysergate (20.6 mg, 0.073 mmol, **24a**:**b** = 35:65) in EtOH (0.68 mL) was added 1N NaOH (0.68 mL). The reaction mixture was stirred at 35 °C for 2 h. 0.1N HCl solution was used to carefully adjust the pH to 6.2 and stirred for further 2 h at 0 °C while a solid material was formed. The precipitate was filtered off and washed with cold water and acetone to give **1** as a pale brown solid (10.6 mg, 54% yield). The IR,

^{1}H NMR and ^{13}C NMR spectra were in agreement with those reported by Hendrickson and Wang [15] and Szántay [16]: IR (neat): 3240 (OH), 1589 (C=O), ^{1}H NMR (500 MHz, C_5D_5N) δ 2.53 (s, 3H), 2.88–2.96 (m, 2H), 3.27–3.33 (m, 1H), 3.53 (dd, J = 11.2, 5.4 Hz, 1H), 3.64 (dd, J = 14.6, 5.4 Hz, 1H), 4.03–4.08 (m, 1H), 7.20–7.26 (m, 2H), 7.30 (dd, J = 8.0, 8.0 Hz, 1H), 7.43 (d, J = 8.0 Hz, 1H), 7.45 (d, J = 8.0 Hz, 1H), 11.68 (s, 1H); ^{1}H NMR (500 MHz, $(CD_3)_2SO$) δ 2.47 (s, 3H), 2.48–2.51 (m, 2H), 2.96–3.02 (m, 1H), 3.13 (dd, J = 11.5, 5.2 Hz, 1H), 3.46 (dd, J = 14.6, 5.4 Hz, 1H), 3.47–3.52 (m, 1H), 6.47 (br s, 1H), 7.01–7.08 (m, 3H), 7.18 (d, J = 7.4 Hz, 1H), 10.70 (br s, 1H); ^{13}C NMR (125 MHz, C_5D_5N) δ 27.8, 43.2, 43.9, 56.0, 63.7, 110.4, 110.5, 112.2, 119.8, 120.1, 127.3, 128.8, 135.8, 136.7, 175.0 (one of the sp^2 carbons was overlapped with C_5D_5N solvent peaks); ^{13}C NMR [125 MHz, $(CD_3)_2SO$] δ 26.6, 41.7, 43.2, 54.6, 62.5, 108.8, 109.9, 111.0, 118.7, 119.3, 122.3, 125.9, 127.3, 133.8, 135.4, 173.4; HRMS (FAB) calcd $C_{16}H_{17}N_2O_2$: $[M{-}H]^-$, 269.1290; found: 269.1289.

References

1. Jacobs W, Craig L (1934) J Biol Chem 104:547–551
2. Stoll A, Hofmann A, Troxler F (1949) Helv Chim Acta 32:506–521
3. Agurell S (1965) Acta Pharm Suecica 2:357–374
4. Agurell S (1966) Acta Pharm Suecica 3:7–10
5. Ninomiya I, Kiguchi T (1990) In: Brossi A (ed) The Alkaloids, vol 38. Academic Press, San Diego, pp 1–156
6. Somei M, Yokoyama Y, Murakami Y, Ninomiya I, Kiguchi T, Naito T (2000) In: Cordell GA (ed) The Alkaloids, vol 54. Academic Press, San Diego, pp 191–257
7. Kornfeld EC, Fornefeld EJ, Kline GB, Mann MJ, Morrison DE, Jones RG, Woodward RB (1956) J Am Chem Soc 78:3087–3114
8. Julia M, LeGoffic F, Igolen J, Baillarge M (1969) Tetrahedron Lett 10:1569–1571
9. Armstrong VW, Coulton S, Ramage R (1976) Tetrahedron Lett 17:4311–4314
10. Oppolzer W, Francotte E, Bättig K (1981) Helv Chim Acta 64:478–481
11. Rebek J Jr, Tai DF (1983) Tetrahedron Lett 24:859–860
12. Kiguchi T, Hashimoto C, Naito T, Ninomiya I (1982) Heterocycles 19:2279–2282
13. Kurihara T, Terada T, Yoneda R (1986) Chem Pharm Bull 34:442–443
14. Cacchi S, Ciattini PG, Morera E, Ortar G (1988) Tetrahedron Lett 29:3117–3120
15. Hendrickson JB, Wang J (2004) Org Lett 6:3–5
16. Moldvai I, Temesvári-Major E, Incze M, Szentirmay É, Gács-Baitz E, Szántay C (2004) J Org Chem 69:5993–6000
17. Inoue T, Yokoshima S, Fukuyama T (2009) Heterocycles 79:373–378
18. Kurosawa T, Isomura M, Tokuyama H, Fukuyama T (2009) Synlett 775–777
19. Kiguchi T, Hashimoto C, Ninomiya I (1985) Heterocycles 23:1377–1380
20. Ninomiya I, Hashimoto C, Kiguchi T, Naito T, Barton DHR, Lusinchi X, Milliet P (1990) J Chem Soc Perkin Trans 1:707–713
21. Deck JA, Martin SF (2010) Org Lett 12:2610–2613
22. Allen MS, Hamaker LK, LaLoggia AJ, Cook JM (1992) Synth Commun 22:2077–2102
23. Johnson WS, Frei B, Gopalan AS (1981) J Org Chem 46:1512–1513
24. Anderson NH, Denniston AD, McCrae DA (1982) J Org Chem 47:1145–1146
25. Corey EJ, Erickson BW (1971) J Org Chem 36:3553–3560
26. Ireland RE, Wipf P, Xiang JN (1991) J Org Chem 56:3572–3582

27. Trost BM, Dong G, Vance JA (2007) J Am Chem Soc 129:4540–4541
28. Gockel B, Krause N (2006) Org Lett 8:4485–4488
29. Sherry BD, Toste FD (2004) J Am Chem Soc 126:15978–15979
30. Bach T, Kather K (1996) J Org Chem 61:7642–7643
31. Ballabio M, Sbraletta P, Mantegani S, Brambilla E (1992) Tetrahedron 48:4555–4566
32. Lauchli R, Shea KJ (2006) Org Lett 8:5287–5289
33. Hong S, Yang J, Weinreb SM (2006) J Org Chem 71:2078–2089

Chapter 5
Total Synthesis of (+)-Lysergic Acid, (+)-Lysergol, and (+)-Isolysergol

Abstract Enantioselective total synthesis of the biologically important indole alkaloids, (+)-lysergol, (+)-isolysergol and (+)-lysergic acid is described. Key features of these total synthesis include: (1) a facile synthesis of a chiral 1,3-amino alcohol via the Pd(0) and In(I)-mediated reductive coupling reaction between L-serine-derived 2-ethynylaziridine and formaldehyde; (2) the Cr(II)/Ni(0)-mediated Nozaki–Hiyama–Kishi (NHK) reaction of an indole-3-acetaldehyde with iodoalkyne; and (3) Pd(0)-catalyzed domino cyclization of an amino allene bearing a bromoindolyl group. This domino cyclization enabled direct construction of the C/D ring system of the ergot alkaloids skeleton as well as the creation of the C5 stereogenic center with transfer of the allenic axial chirality to the central chirality.

As described in Chap. 4, ergot alkaloids, particularly lysergic acid (**1**), have attracted considerable interest from the synthetic community, because of their biological importance as well as structural appeal (Fig. 5.1) (For enantioselective synthesis of lysergic acid, see: [1–3]; For enantioselective synthesis of isolysergol, see: [4]). The pivotal steps toward the total synthesis are the construction of the C/D ring system controlling the stereochemistry at C5. Despite intensive synthetic investigations, there are only three asymmetric syntheses reported: Szántay in 2004 [1], and Fukuyama in 2009 [2, 3]. The former involves optical resolution of the tetracyclic indole intermediate with L-tartaric acid, and the latter two utilize a stepwise or double cyclization strategy for the construction of the B/C ring.

Cyclization reaction of a functionalized allenes is a valuable method for the synthesis of chiral cyclic compounds. It is well known that the axial chirality of allenes is stereospecifically transferred into the new stereogenic centers in the cases of Ag (For recent reviews on Ag-mediated axis-to-center chirality transfer of amino allenes, see: [5–7]), Au (For recent reviews on Au-mediated axis-to-center chirality transfer of amino allenes, see: [8–14]), organolanthanide [15, 16] or K_2CO_3 [17]-mediated cyclization of amino allenes (Scheme 5.1, Eq. 1). In contrast, when using palladium-catalyzed cyclization with aryl halides, prediction of the product

S. Inuki, *Total Synthesis of Bioactive Natural Products by Palladium-Catalyzed Domino Cyclization of Allenes and Related Compounds*, Springer Theses,
DOI: 10.1007/978-4-431-54043-4_5, © Springer 2012

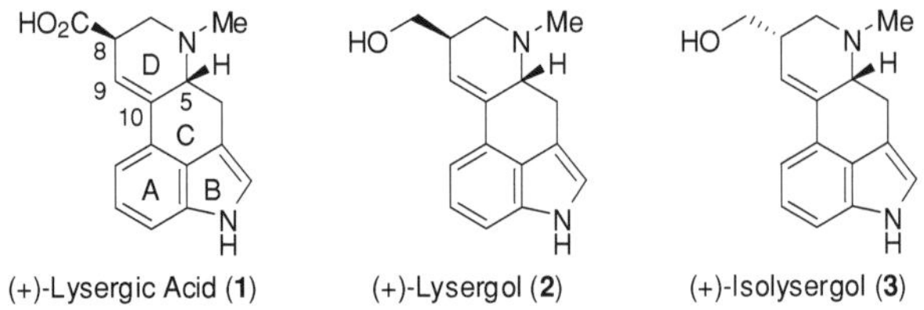

(+)-Lysergic Acid (**1**) (+)-Lysergol (**2**) (+)-Isolysergol (**3**)

Fig. 5.1 Indole alkaloids of the ergot family

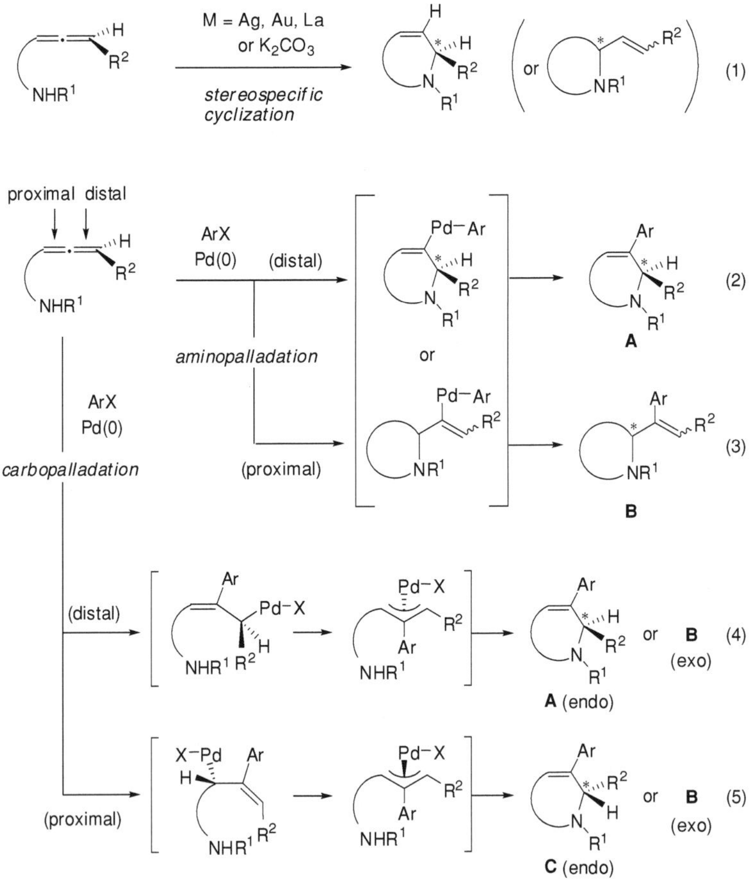

Scheme 5.1 Product distribution of transition metal-mediated cyclization of allenes allenes

distribution including stereo- and regioisomers is more difficult, because these types of reactions may proceed through two competing pathways (Scheme 5.1): the aminopalladation pathway, where the arylpalladium halide would activate the distal

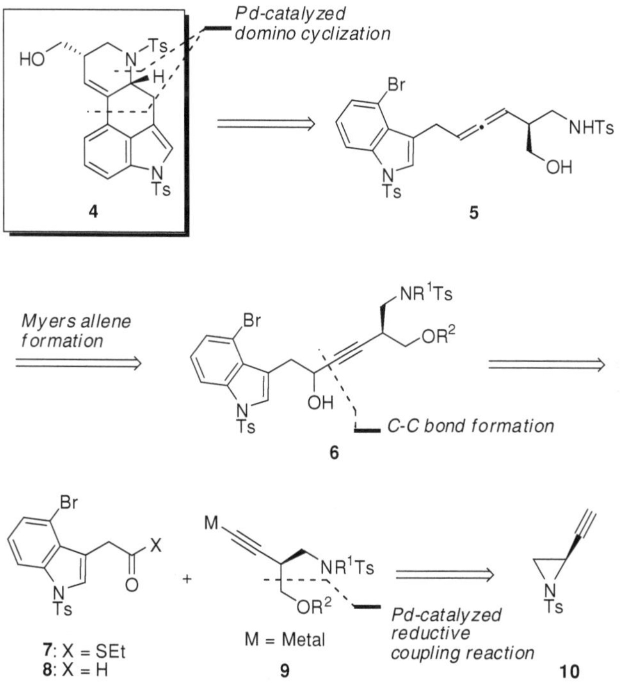

Scheme 5.2 Retrosynthetic analysis of the ergot alkaloid core structure **4**

double bond from the less hindered side (Eq. 2), affords the *endo*-type cyclization product **A** stereospecifically through reductive elimination, while the reaction at the proximal double bond gives its regioisomer **B** (Eq. 3). On the other hand, carbopalladation onto the distal double bond from the less hindered side (Eq. 4) followed by *anti*-cyclization of the η^3-allylpalladium intermediate by the nitrogen nucleophile would give the *endo*-cyclization product **A**, which has the same configuration as the product formed by distal bond aminopalladation (Eq. 2). However, the reaction at the proximal double bond would provide the *endo*-cyclization product **C** (Eq. 5), which has the opposite configuration to the distal aminopalladation product **A** (Eq. 2). Consideration of the *exo*-type cyclization to produce **B** from the η^3-allylpalladium intermediate will make the prediction more complicated.

Based on the synthetic studies on the racemic ergot alkaloids as described in Chap. 4, the author expected a palladium-catalyzed domino cyclization of chiral amino allenes **5** bearing a protected 4-bromoindol-3-yl group and a free hydroxy group to provide the direct construction of the desired chiral ergot alkaloids skeleton (Scheme 5.2). This bis-cyclization would allow the simultaneous construction of the C/D ring system and the creation of the C5 chiral center. The challenges in this domino cyclization are transfer of an axial chirality in the starting allene to the central chirality at C5. Enantioselective total syntheses of lysergic acid (**1**), lysergol (**2**) and isolysergol (**3**) based on this strategy are also presented.

Table 5.1 Reductive coupling reaction of 2-ethynylaziridine **10** with formaldehyde[a]

Pd(PPh3)4, InI, aldehyde, solvent: **10** (97% ee) → **11**

Entry	Aldehyde	Solvent	Additive (equiv.)	Yield (%)[b]	ee (%)[c]
1	$(CH_2O)_n$	THF:HMPA (4:1)	H_2O (1.0)	78	96
2	$(CH_2O)_n$	THF:DMPU (4:1)	H_2O (1.0)	77	83
3	$(CH_2O)_n$	THF:H_2O (1:1)	–	50	92
4	$(CH_2O)_n$	DMF:H_2O (1:1)	–	15	58
5	$(CH_2O)_n$	THF	–	ca. 42	91
6	formalin	THF:HMPA (4:1)	–	83	97
7[d]	formalin	THF:HMPA (4:1)	–	88 (70[e])	97 (99[e])

[a] Reactions were carried out using the aziridine **10** (97% ee) with $Pd(PPh_3)_4$ (5 mol %), InI (1.3 equiv.) and aldehyde (2.0 equiv.) for 1.5–4 h
[b] Isolated yields
[c] Determined by chiral HPLC (OD-H) analysis
[d] Reaction was carried out with $Pd(PPh_3)_4$ (3 mol %) and InI (1.2 equiv.) on a 4 g scale
[e] After single recrystallization

Retrosynthetic analysis of the amino allene **5** is shown in Scheme 5.2.[1] The author planned to synthesize both diastereomeric amino allenes **5** in order to examine the difference in reactivity between these isomers. The chiral allene unit of **5** would form from chiral propargyl alcohol **6** using the Myers method [18]. The propargyl alcohol **6** could be obtained by C–C bond formation reaction of thioester **7** or aldehyde **8** with metal acetylide **9**, in combination with asymmetric hydrogenation if necessary. The precursor of the acetylide **9** can be accessed from L-serine-derived chiral 2-ethynylaziridine **10** by a reductive coupling reaction with formaldehyde in the presence of $Pd(PPh_3)_4$ and InI, as the author's group previously reported [19, 20].

Initially, the author investigated the palladium-catalyzed reductive coupling reaction of ethynylaziridine **10** (Table 5.1). The aziridine **10** was easily prepared in an enantioenriched form (97% ee) by a four-step sequence from the (*S*)-Garner's aldehyde [21, 22] (alkyne formation, deprotection, *N*-tosylation, and aziridine formation), following the author's group reported procedure [23]. The previous study by the author's group revealed that the reductive coupling reaction of 2,3-*cis*- or 2,3-*trans*-2-ethynylaziridines efficiently reacts with alkyl or aryl aldehyde in the presence of InI and a catalytic amount of Pd(0) to produce 2-ethynyl-1, 3-amino alcohols in a highly stereoselective manner (mostly $>$ 99:1) [19, 20]. In the present case, using the aziridine **10** lacking the 3-substituent required careful

[1] The author planned to develop a new synthetic route to the chiral amino allenes **5** because the synthetic route described in Chap. 4 gave the racemic amino allenes of the type **5** in low diastereoselectivities.

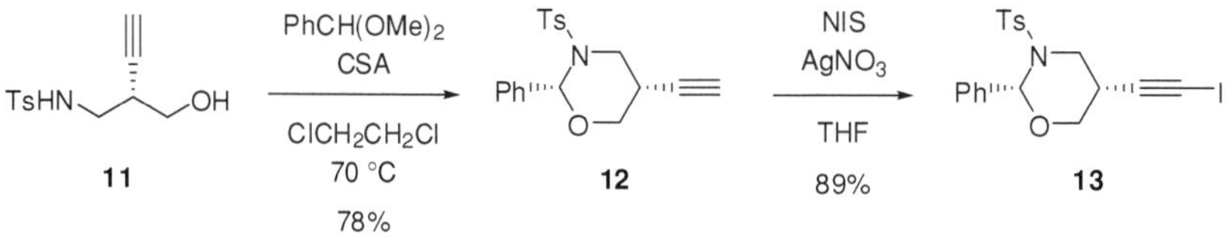

Scheme 5.3 Synthesis of iodoalkyne **13**

investigation, because the stereoselectivity of the reaction would be reflected in the enantiomeric purity of the resulting amino alcohol **11**. Treatment of **10** with $(CH_2O)_n$, $Pd(PPh_3)_4$ (5 mol%), and InI in THF/HMPA (standard conditions for the preparation and addition of the allenylindium reagents) [19, 20] produced the desired 1,3-amino alcohol **11** (96% ee) in 78% yield (entry 1). Changing the reaction solvent from THF/HMPA to THF/DMPU, THF/H_2O, DMF/H_2O or THF only decreased the optical purity of the desired product **11** without improving the yield (entries 2–5). Use of formalin instead of $(CH_2O)_n$ afforded the desired product in a higher yield (83%) in good stereoselectivity (97% ee, entry 6). Conducting the reaction on a 4 g scale also gave the desired product in satisfactory yield (88%, entry 7), and the enantiomerically pure alcohol **11** was obtained after single recrystallization. Protection of the 1,3-amino alcohol **11** as benzylidene acetal provided the desired alkyne **12**,[2] which was allowed to react with NIS and $AgNO_3$ to give the corresponding iodoalkyne **13** (Scheme 5.3) [24].

The author next examined the preparation of ynone **16** by palladium-mediated coupling of a thioester with an alkyne, which is known to proceed under mild conditions (Scheme 5.4) [25]. The requisite thioester **7** for the coupling reaction was prepared by the hydrolysis of a known nitrile **14** [26], thioesterification, and *N*-protection of indole. Unfortunately, the reaction of **7** with the alkyne **12** in the presence of $Pd_2(dba)_3{\cdot}CHCl_3$ (5 mol %), P(2-furyl)$_3$, and CuI in DMF/Et_3N at 50 °C afforded the desired product **16** in low yield (ca. 37%) along with several unidentified side products.

The author next investigated the cross-coupling reaction of the alkyne **12** or iodoalkyne **13** with (4-bromoindol-3-yl)acetaldehyde **8** (Table 5.2), which was prepared from commercially available 4-bromoindole **17** as follows (Scheme 5.5).

[2] The relative configuration of **12** was determined by ^{1}H NOE analysis.

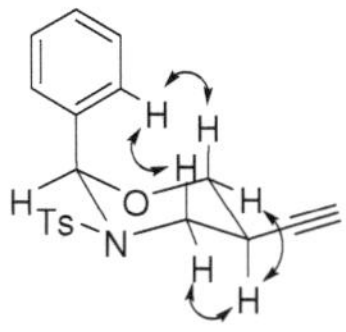

Selected NOE cross peaks for **12**

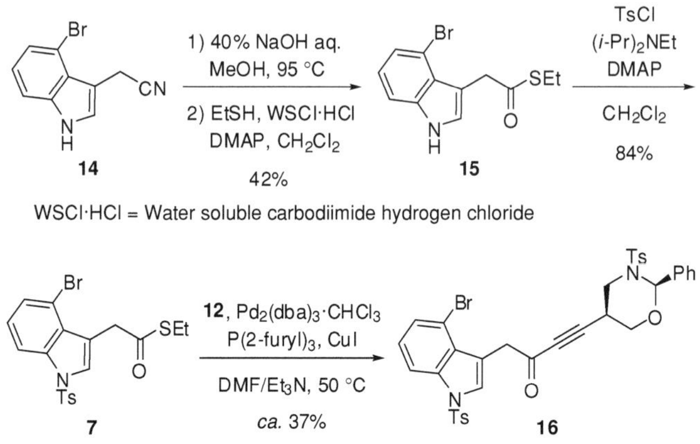

Scheme 5.4 Synthesis of ynone **16** by palladium-catalyzed coupling of thioester **15** with alkyne **12**

Table 5.2 Cross-coupling reaction of aldehyde **8** and alkyne **12** or iodoalkyne **13**

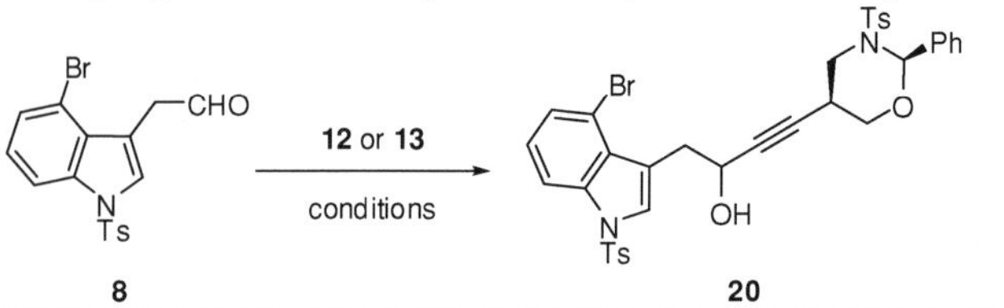

Entry	Substr.	Conditions	Yield (%)[a]	dr[b]
1	**12**	*n*-BuLi, THF, −78 °C	50	1:1
2	**12**	*n*-BuLi, $CeCl_3$, THF, −78 °C	67	1:1
3	**12**	$InBr_3$, (*R*)-BINOL, Cy_2NMe, CH_2Cl_2, 40 °C	ND	–
4	**12**	Et_2Zn, (*S*)-BINOL, Ti(O*i*-Pr)$_4$, toluene/THF, 0 °C	ND	–
5	**13**	$NiCl_2$, $CrCl_2$, THF, 0 °C	90	1:1

[a] Isolated yields
[b] Determined by ^{1}H NMR analysis

3-Allylindole **18** was obtained using palladium-catalyzed C3-selective allylation of indoles with allyl alcohol and triethylborane, reported by Tamaru [27]. *N*-Protection of indole **18** [28] followed by $OsO_4/NaIO_4$-mediated oxidative cleavage of the double bond gave the desired aldehyde **8**. The addition of **12**-derived lithium acetylide with the aldehyde **8** provided the desired propargyl alcohol **20** in a moderate yield (50%, dr = 1:1, Table 5.2, entry 1). Addition of $CeCl_3$ improved the yield to 67% (entry 2) [29]. In contrast, mild conditions using $InBr_3$ [30] or Et_2Zn [31–33] did not afford the desired product, although the starting aldehyde was consumed (entries 3 and 4). Successful cross-coupling was achieved using the Cr(II)/Ni(0)-mediated Nozaki–Hiyama–Kishi (NHK) reaction with **8** and the iodoalkyne **13**, leading to the desired product **20** in 90% yield

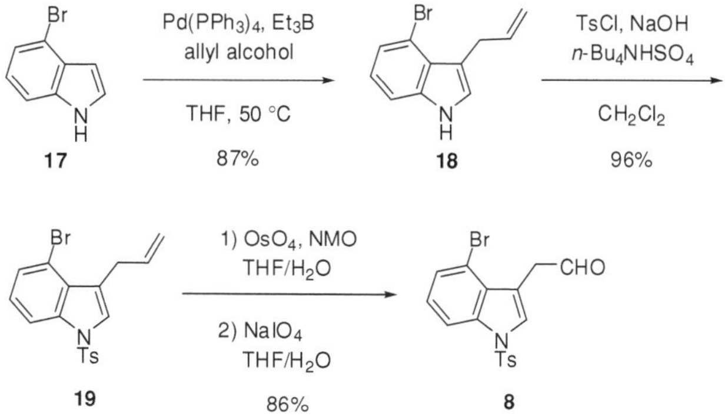
Br
Pd(PPh3)4, Et3B
allyl alcohol
THF, 50 °C
87%
N
H
17
Br
N
H
18
TsCl, NaOH
n-Bu4NHSO4
CH2Cl2
96%
Br
N
Ts
19
1) OsO4, NMO
THF/H2O
2) NaIO4
THF/H2O
86%
Br
CHO
N
Ts
8

Scheme 5.5 Synthesis of aldehyde **8**

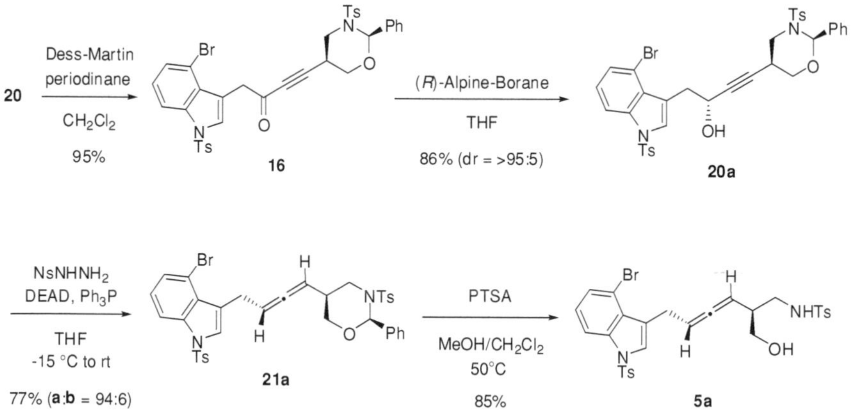
20
Dess-Martin
periodinane
CH2Cl2
95%
Br
N
Ts
O
Ts
N
Ph
O
16
(R)-Alpine-Borane
THF
86% (dr = >95:5)
Br
N
Ts
OH
20a
NsNHNH2
DEAD, Ph3P
THF
-15 °C to rt
77% (a:b = 94:6)
Br
H
H
NTs
O
Ph
21a
PTSA
MeOH/CH2Cl2
50°C
85%
Br
H
H
NHTs
OH
5a

Scheme 5.6 Synthesis of allenic amide **5a**

(dr = 1:1, entry 5) [34–36], (Any further attempt to carry out the asymmetric NHK reaction using chiral sulfonamide ligands failed to produce the desired propargyl alcohol. For asymmetric NHK reactions, see: [37–39]).

With the propargyl alcohol **20** in hand, the author attempted the conversion to each isomer of the requisite allenic amides **5** for the palladium-catalyzed domino cyclization (Scheme 5.6 and 5.7). Dess–Martin oxidation of **20** followed by reduction with (*R*)-Alpine-Borane [40] furnished the desired propargyl alcohol **20a** in 86% yield with high diastereoselectivity (dr = > 95:5, Scheme 5.6) (Use of other reagents for the asymmetric reduction, such as the Noyori's Ru-TsDPEN complex [41] or Me-CBS-catalyst [42–44], led to a decrease in the yield and diastereoselectivity). This alcohol was stereoselectively transformed into the allene **21a** by the Myers method using nosyl hydrazine under Mitsunobu conditions [18]. Subsequent cleavage of the benzylidene group of **21a** with PTSA gave the allenic amide **5a** (dr = 94:6).[3] The diastereomeric allenic amide **5b** (dr = 94:6) was similarly prepared from the same propargyl ketone **16**, via reduction with (*S*)-Alpine-Borane (Scheme 5.7).

The author next examined the construction of the ergot alkaloid skeleton via the palladium-catalyzed domino cyclization of the allenic amides **5** bearing a free hydroxy group (Scheme 5.8). The reaction was conducted using a 94:6 diastereomixture of **5a** and **5b** because of the difficulty in separating each of the diastereomers. Reaction of **5a** with 5 mol % of $Pd(PPh_3)_4$ and K_2CO_3 in DMF at 100 °C (the optimized conditions for the domino cyclization of racemic model substrates as described in Chap. 4) provided the desired product **4** in 76% yield with good diastereoselectivity (**a**:**b** = 92:8).[4] The dihydropyran derivatives (the cyclization by the hydroxy group) and/or the azetidine derivatives (the proximal cyclization by the NHTs group) were not isolated as side products. When the diastereomeric allenic amide **5b** was subjected to the same conditions, the yield and stereoselectivity of the reaction was dramatically reduced (43% yield, **a**:**b** = 31:69). These results show a clear difference in reactivity between the diastereomeric substrates.

A rationale for stereoselectivities of the domino cyclization of internal amino allenes is depicted in Scheme 5.9. This domino cyclization could proceed through two pathways: (1) carbopalladation and (2) aminopalladation. Because of a steric reason, carbopalladation of indolylpalladium(II) bromide, formed in situ by oxidative addition of the bromoindole moiety to Pd(0), would proceed through 6-exo type cyclization as depicted in **D** to generate η^3-allylpalladium complex **E**. The second cyclization by the tosylamide group in an *anti*-manner then gives the minor isomer **4b**. On the other hand, coordination of the indolylpalladium(II) to the allenic moiety would promote *anti*- attack of the tosylamide group as shown in

[3] The relative configuration of allenic amide **5a** was confirmed by comparison with the authentic sample (±)-**5a** prepared from the known allenic amide (±)-**22a**, which, in turn, was obtained through an Au-catalyzed Claisen rearrangement of the corresponding propargyl vinyl ether (see Chap. 4).

Br H NHTs OTIPS H N Ts (±)-**22a** —TBAF, THF, 91%→ (±)-**5a**

[4] The relative configuration of **4a** was confirmed by comparison with the authentic sample prepared from the known compound (±)-**23a** (see Chap. 4).

TIPSO N Ts H N Ts (±)-**23a** —TBAF, THF, quant.→ (±)-**4a**

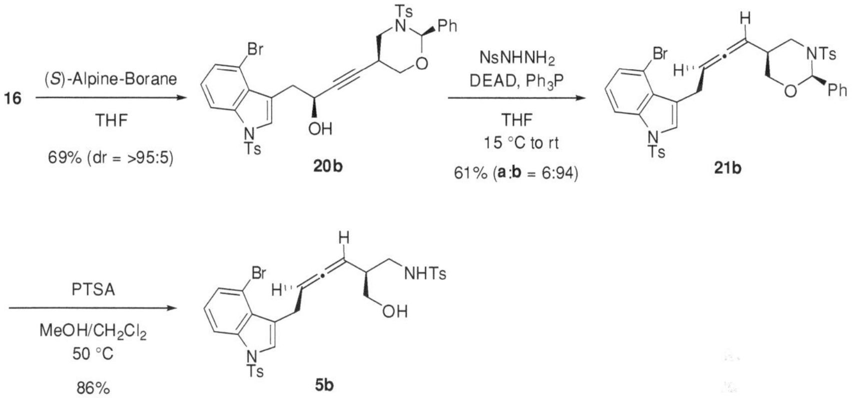

Scheme 5.7 Synthesis of allenic amide **5b**

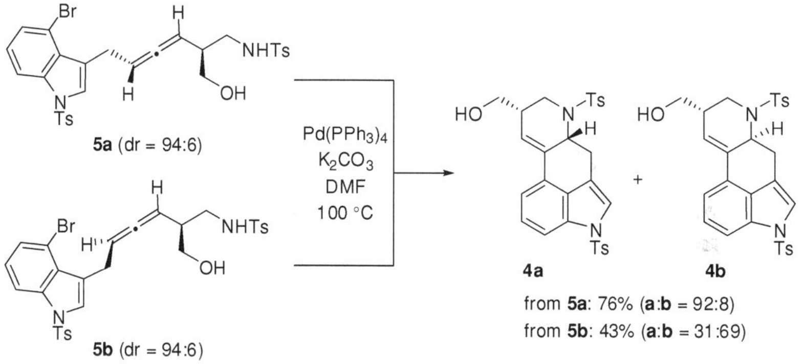

Scheme 5.8 Palladium-catalyzed domino cyclization of allenic amides **5a** and **5b**

F (aminopalladation pathway) to give a palladacycle **G**, which gives the isomer **4a** by reductive elimination. Predominant formation of **4a** can be rationalized by considering the strained bicyclic structure **D** in the carbopalladation step.

Next, the low reactivity and selectivity of diastereomeric allenic amide **5b** in comparison with those of **5a** (Scheme 5.8) can be explained in Scheme 5.10. The cyclization of **5b** would also proceed mainly through reactive conformer *epi*-**F** (aminopalladation pathway) to give **4b**. However, unfavorable steric interaction between the tosylamide group and the methylene protons both located on the same side destabilizes this conformer, which would decrease reactivity of **5b** toward aminopalladation via *epi*-**F**.[5] Thus, the cyclization reaction of the allenic amide **5b**

[5] The author cannot rule out other factors for rationalization of the observed selectivities. For example, the reactive conformer as depicted in **F** might have better orbital alignment for anti-addition of the amine nucleophile to the allenic moiety activated by Pd(II) than in *epi*-**F**, thus leading to a selective formation of the desired product **4a**.

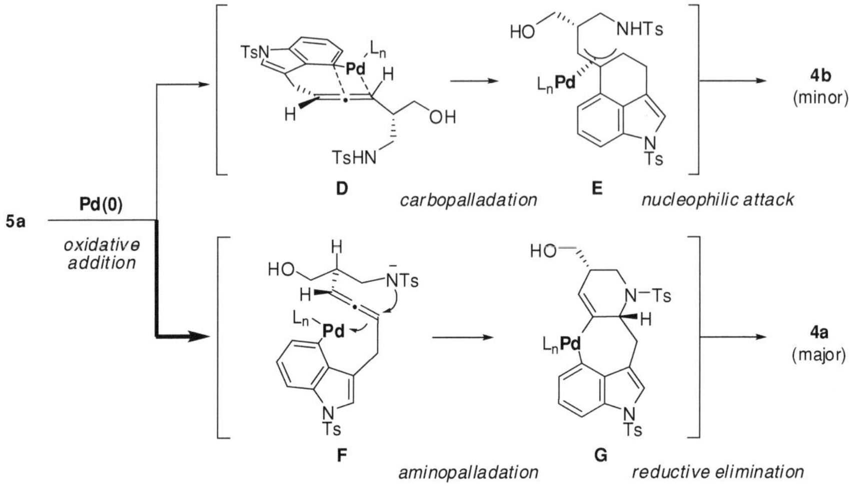

Scheme 5.9 Proposed mechanism for domino cyclization of **5a**

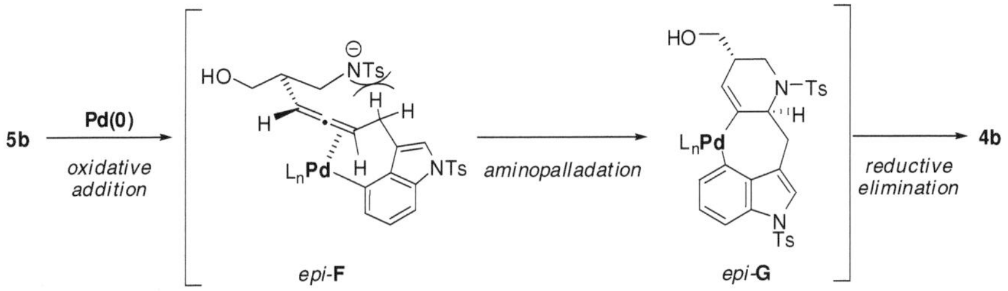

Scheme 5.10 Proposed mechanism for domino cyclization of **5b**

may partially involve aminopalladation through other conformers or the competing carbopalladation pathway.

With the ergot derivatives **4** with all the requisite functionalities in hand, the author then focused on the total synthesis of isolysergol (**3**), lysergol (**2**) and lysergic acid (**1**) on the basis of the synthetic studies on the racemic ergot alkaloids as described in Chap. 4 (Scheme 5.11 and 5.12). Cleavage of the tosyl groups of **4a** with sodium naphthalenide and subsequent *N*-methylation led to (+)-isolysergol (**3**) in 46% yield (99% ee, Chiralcel OD-H) [4] (Scheme 5.11). Oxidation of the primary alcohol of **4a** with the Dess–Martin reagent[6] and $NaClO_2$ followed by esterification with $TMSCHN_2$ gave the corresponding methyl ester **24a** (64%,

[6] The reproducibility of the oxidation reaction was significantly dependent on the purity of the Dess-Martin reagent.

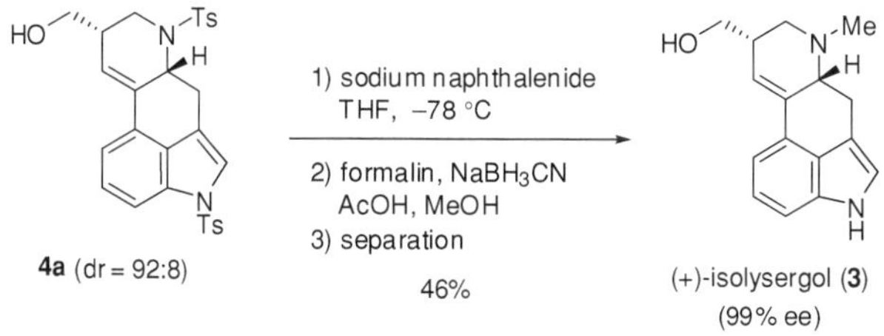

Scheme 5.11 Total synthesis of isolysergol (**3**)

3 steps), after separation of the diastereomers (Scheme 5.12).[7] Cleavage of two tosyl groups with sodium naphthalenide and subsequent *N*-methylation led to a diastereomixture of methyl isolysergate **25a** and lysergate **25b** (65%, **a**:**b** = 33:67). By reduction of **25** (**a**:**b** = 33:67) with $LiAlH_4$, (+)-lysergol (**2**) was obtained in 49% yield (98% ee, Chiralcel OD-H), along with (+)-isolysergol (**3**) (24%) [45]. Finally, hydrolysis of **25** (**a**:**b** = 33:67) with NaOH accompanying isomerization to the natural isomer [1], furnished (+)-lysergic acid (**1**) in 54% yield (96% ee, Chiralcel OD-H after methylation with $TMSCHN_2$) (Lysergic acid is known to racemize under harsh basic conditions [$Ba(OH)_2$ aq., sealed-tube, 150 °C, 4 h], see: [46, 47]).[8] All the spectroscopic data were in agreement with those of natural and synthetic (+)-lysergic acid, (+)-lysergol and (+)-isolysergol reported in the literature [1, 4].

In conclusion, the enantioselective total synthesis of (+)-lysergol, (+)-isolysergol and (+)-lysergic acid has been accomplished. (+)-Lysergic acid was prepared in 15 steps from the known ethynylaziridine (4.0% overall yield; 19 steps, 1.1% overall yield from the Garner's aldehyde). The author's synthesis highlights a strategy for constructing the C/D ring system of the core structure of ergot alkaloids based on palladium-catalyzed domino cyclization of chiral amino allene, which allows the creation of the stereochemistry at C5 by transfer of the axial chirality of allene to the central chirality. Other key features of the syntheses include the Pd(0)/In(I)-mediated reductive coupling reaction of chiral

[7] Separation of the diastereomer at this step is important for the preparation of lysergic acid (**1**) in high ee, because the transformation of **25** to **1** accompanying isomerization relies on the chirality at C-5.

[8] The optical purity of lysergic acid was confirmed by derivatization to methyl isolysergate **25a** and lysergate **25b** and their chiral HPLC analyses.

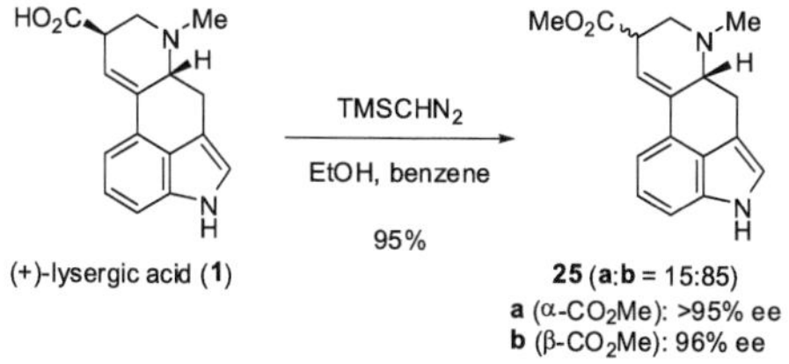

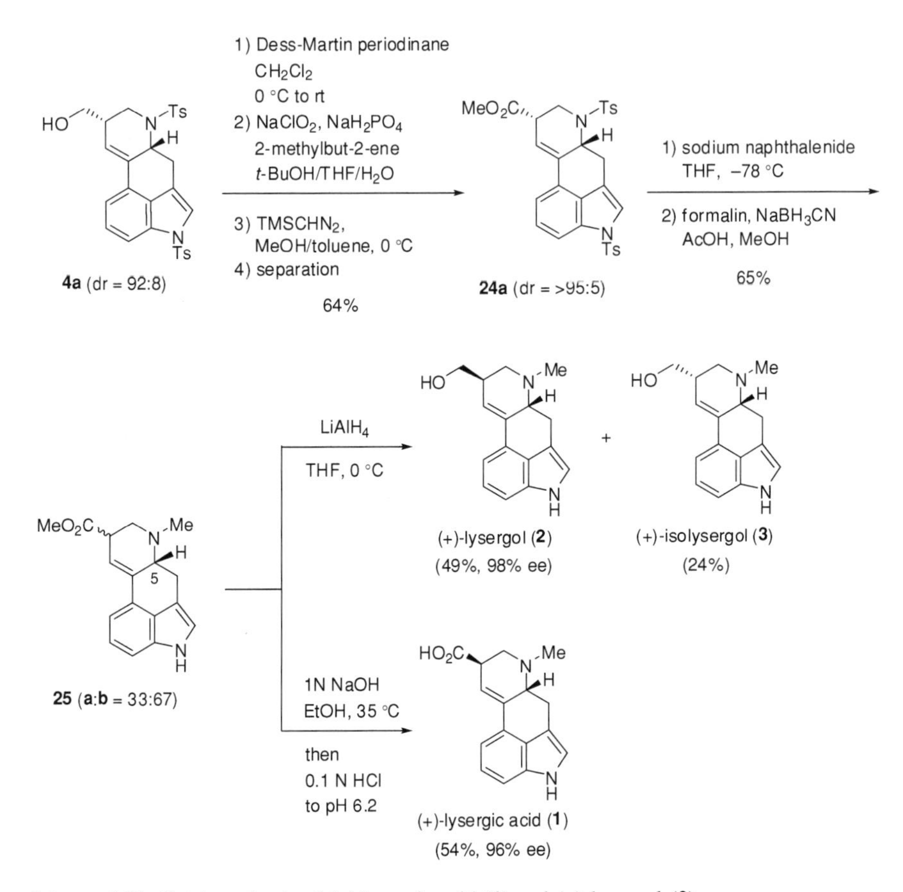

Scheme 5.12 Total synthesis of (+)-lysergic acid (**1**) and (+)-lysergol (**2**)

2-ethynylaziridine with formaldehyde, and the Cr(II)/Ni(0)-mediated Nozaki–Hiyama–Kishi (NHK) reaction of indole-3-acetaldehyde with iodoalkyne.

5.1 Experimental Section

5.1.1 General Methods

All moisture-sensitive reactions were performed using syringe-septum cap techniques under an argon atmosphere and all glassware was dried in an oven at 80 °C for 2 h prior to use. Reactions at −78 °C employed a CO_2–MeOH bath. Melting points were measured by a hot stage melting point apparatus (uncorrected). Chemical shifts are reported in δ (ppm) relative to TMS in $CDCl_3$ as internal standard (^{1}H NMR) or the residual $CHCl_3$ signal (^{13}C NMR). ^{1}H NMR spectra are tabulated as follows: chemical shift, multiplicity (b = broad, s = singlet, d = doublet, t = triplet, q = quartet, m = multiplet), number of protons, and coupling constant(s).

5.1.2 (S)-N-[2-(Hydroxymethyl)but-3-ynyl]-4-methylbenzenesulfonamide (11)

To a stirred mixture of aziridine **10** (4.00 g, 18.1 mmol, 97% ee) in THF/HMPA (150 mL, 4:1) were added $Pd(PPh_3)_4$ (627 mg, 0.54 mmol), InI (5.25 g, 21.7 mmol) and formalin (2.7 mL, 36.2 mmol) at room temperature under argon (Table 5.1, Entry 7). The mixture was stirred for 2.5 h at this temperature, and filtered through a short pad of silica gel with EtOAc to give a crude **11**. The residue was dissolved in Et_2O and washed with H_2O, brine and dried over $MgSO_4$. The filtrate was concentrated under reduced pressure to give a yellow oil, which was purified by flash chromatography over silica gel with *n*-hexane–EtOAc (2:1) to give **11** as a yellow solid {4.01 g, 88% yield, 97% ee [HPLC, Chiralcel-OD column eluting with 90:10 hexane/EtOH at 0.5 mL/min, t_1 = 26.10 min (major isomer), t_2 = 30.67 min (minor isomer)]}. Recrystallization from *n*-hexane–EtOAc gave pure **11** (3.23 g, 99% ee) as colorless crystals: mp 86–87 °C; $[\alpha]_D^{26}$ −14.8 (*c* 1.06, $CHCl_3$); IR (neat): 3289 (OH), 1327 (NSO_2), 1158 (NSO_2); ^{1}H NMR (500 MHz, $CDCl_3$) δ 2.15 (d, J = 2.3 Hz, 1H), 2.39 (t, J = 6.6 Hz, 1H), 2.43 (s, 3H), 2.69–2.77 (m, 1H), 3.14 (ddd, J = 12.6, 6.6, 5.7 Hz, 1H), 3.21 (ddd, J = 12.6, 6.6, 5.1 Hz, 1H), 3.72–3.77 (m, 2H), 5.07 (t, J = 6.6 Hz, 1H), 7.32 (d, J = 8.0 Hz, 2H), 7.75 (d, J = 8.0 Hz, 2H); ^{13}C NMR (125 MHz, $CDCl_3$) δ 21.5, 34.7, 43.2, 62.2, 72.7, 81.3, 127.0 (2C), 129.8 (2C), 136.8, 143.7. *Anal.* Calcd for $C_{12}H_{15}NO_3S$: C, 56.90; H, 5.97; N, 5.53. Found: C, 56.74; H, 5.84; N, 5.50.

5.1.3 (2R,5S)-5-Ethynyl-2-phenyl-3-tosyl-1,3-oxazinane (12)

To a stirred mixture of **11** (1.70 g, 6.70 mmol) and $PhCH(OMe)_2$ (2.0 mL, 13.4 mmol) in $ClCH_2CH_2Cl$ (40 mL) was added camphor-10-sulfonic acid (156 mg, 0.67 mmol) at room temperature. The mixture was stirred for 14 h at 70 °C and quenched with saturated $NaHCO_3$. The mixture was diluted with EtOAc. The organic phase was separated and washed with H_2O, brine and dried over $MgSO_4$. The filtrate was concentrated under reduced pressure to give an oily residue, which was purified by flash chromatography over silica gel with *n*-hexane–EtOAc (10:1) to give **12** as a white solid (1.78 g, 78% yield). Recrystallization from *n*-hexane–EtOAc gave pure **12** as colorless crystals: mp 125–126 °C; $[\alpha]_D^{27}$ −57.3 (*c* 0.93, $CHCl_3$); IR (neat): 1348 (NSO_2), 1167 (NSO_2);

^{1}H NMR (500 MHz, $CDCl_3$) δ 1.96 (d, J = 2.3 Hz, 1H), 2.25–2.35 (m, 1H), 2.47 (s, 3H), 3.18 (dd, J = 14.9, 12.0 Hz, 1H), 3.55 (dd, J = 11.6, 11.2 Hz, 1H), 3.70 (dd, J = 11.6, 4.6 Hz, 1H), 4.01 (dd, J = 14.9, 4.6 Hz, 1H), 6.70 (s, 1H), 7.34–7.38 (m, 1H), 7.38 (d, J = 8.0 Hz, 2H), 7.41–7.48 (m, 4H), 7.87 (d, J = 8.0 Hz, 2H); ^{1}H NMR (500 MHz, C_6D_6) δ 1.45 (d, J = 2.9 Hz, 1H), 1.83 (s, 3H), 2.32–2.39 (m, 1H), 3.19 (dd, J = 14.9, 11.7 Hz, 1H), 3.44 (dd, J = 11.3, 5.0 Hz, 1H), 3.48 (dd, J = 11.3, 10.8 Hz, 1H), 4.22 (dd, J = 14.9, 4.6 Hz, 1H), 6.72 (d, J = 8.0 Hz, 2H), 6.92 (s, 1H), 7.04 (t, J = 7.4 Hz, 1H), 7.10 (dd, J = 7.4, 7.4 Hz, 2H), 7.44 (d, J = 7.4 Hz, 2H), 7.75 (d, J = 8.0 Hz, 2H); ^{13}C NMR (125 MHz, $CDCl_3$) δ 21.6, 25.5, 43.9, 63.2, 72.0, 79.8, 83.0, 127.0 (2C), 127.5 (2C), 128.5, 129.1 (2C), 130.0 (2C), 135.0, 137.5, 144.0. *Anal.* Calcd for $C_{19}H_{19}NO_3S$: C, 66.84; H, 5.61; N, 4.10. Found: C, 66.91; H, 5.71; N, 4.04.

5.1.4 (2R,5S)-5-(Iodoethynyl)-2-phenyl-3-tosyl-1,3-oxazinane (13)

To a stirred mixture of **12** (100 mg, 0.29 mmol) in THF (1.0 mL) were added *N*-iodosuccinimide (98.8 mg, 0.44 mmol) and $AgNO_3$ (7.39 mg, 0.044 mmol) at room temperature. The mixture was stirred for 2 h at this temperature and quenched with ice-cold H_2O. The whole was extracted with EtOAc. The extract was washed with saturated $Na_2S_2O_3$, H_2O, brine and dried over $MgSO_4$, and concentrated under pressure to give a white solid, which was purified by column chromatography over silica gel with *n*-hexane–EtOAc (10:1) to give **13** (121 mg, 89% yield). Recrystallization from benzene gave pure **13** as colorless crystals: mp 75–76 °C; $[\alpha]_D^{27}$ −108.9 (*c* 1.00, $CHCl_3$); IR (neat): 1343 (NSO_2), 1165 (NSO_2); ^{1}H NMR (500 MHz, $CDCl_3$) δ 2.39–2.47 (m, 1H), 2.47 (s, 3H), 3.17 (dd, J = 14.9, 12.0 Hz, 1H), 3.54 (dd, J = 11.5, 10.9 Hz, 1H), 3.65–3.72 (m, 1H), 3.99 (dd, J = 14.9, 4.6 Hz, 1H), 6.68 (s, 1H), 7.34–7.37 (m, 1H), 7.37 (d, J = 8.0 Hz, 2H), 7.41–7.47 (m, 4H), 7.86 (d, J = 8.0 Hz, 2H); ^{13}C NMR (125 MHz, $CDCl_3$) δ –0.7, 21.6, 27.7, 44.0, 63.2, 83.0, 90.0, 127.1 (2C), 127.5 (2C), 128.6, 129.2 (2C), 130.1 (2C), 134.9, 137.4, 144.1. *Anal.* Calcd for $C_{19}H_{18}NO_3S+0.75C_6H_6$: C, 53.67; H, 4.31; N, 2.66. Found: C, 53.77; H, 4.31; N, 2.57.

5.1.5 S-Ethyl 2-(4-Bromo-1H-indol-3-yl)ethanethioate (15)

The hydrolysis of 4-bromo-3-indoleacetonitrile **14** was carried out according to the method of Somei [26]. To a stirred solution of the 4-bromo-3-indoleacetonitrile **14** (6.28 g, 26.8 mmol) in MeOH (200 mL) was added 40% aq. NaOH (200 ml), and the mixture was stirred for 4.5 h at 95 °C. MeOH was removed under reduced pressure. After addition of brine, the whole was made acidic by adding conc. HCl, and the extracted with CH_2Cl_2. The extract was washed with brine, dried over

$MgSO_4$. Concentration of the filtrate under reduced pressure to give 3.5 g of crude indoleacetic acid, which was used without further purification. To a stirred solution of the indoleacetic acid in CH_2Cl_2 (160 mL) was added DMAP (84.3 mg, 0.69 mmol), EtSH (3.23 mL, 55.3 mmol) and WSCI· HCl (3.21 g, 16.7 mmol) at 0 °C. After stirring for 2.5 h at room temperature, H_2O was added, and the mixture was concentrated under reduced pressure. The residue was diluted with EtOAc. The extract was washed with H_2O, brine and dried over $MgSO_4$, and concentrated under pressure to give a white solid, which was purified by column chromatography over silica gel with *n*-hexane–EtOAc (5:1) to give **15** (3.38 g, 42% yield). Recrystallization from *n*-hexane–EtOAc gave pure **15** as colorless crystals: mp 96–97 °C; IR (neat): 3378 (NH), 1658 (C=O); ^{1}H NMR (400 MHz, $CDCl_3$) δ 1.23 (t, $J = 7.4$ Hz, 3H), 2.87 (q, $J = 7.4$ Hz, 2H), 4.22 (s, 2H), 7.00 (dd, $J = 7.8$, 7.8 Hz, 1H), 7.11 (d, $J = 2.4$ Hz, 1H), 7.27 (d, $J = 7.8$ Hz, 1H), 7.27 (d, $J = 7.8$ Hz, 1H), 8.31 (s, 1H); ^{13}C NMR (100 MHz, $CDCl_3$) δ 14.6, 23.5, 41.1, 108.9, 110.7, 114.2, 123.2, 124.3, 125.5, 126.1, 137.4, 199.8. *Anal.* Calcd for $C_{12}H_{12}BrNOS$: C, 48.33; H, 4.06; N, 4.70. Found: C, 48.46; H, 4.09; N, 4.69.

5.1.6 S-Ethyl 2-(4-Bromo-1-tosyl-1H-indol -3-yl)ethanethioate (7)

To a stirred solution of thioester **15** (3.38 g, 11.4 mmol) in CH_2Cl_2 (50 mL) were added TsCl (4.77 g, 25.0 mmol), (*i*-Pr)$_2$NEt (4.36 mL, 25.0 mmol) and DMAP (278 mg, 2.28 mmol) at 0 °C. The mixture was stirred for 5 h at this temperature and quenched with saturated NH_4Cl. The whole was extracted with EtOAc. The extract was washed with saturated $NaHCO_3$, brine and dried over $MgSO_4$, and concentrated under pressure to give a white solid, which was purified by column chromatography over silica gel with *n*-hexane–EtOAc (10:1) to give **7** (4.35 g, 84% yield). Recrystallization from *n*-hexane–EtOAc gave pure **7** as colorless crystals: mp 99–100 °C; IR (neat): 1680 (C=O), 1372 (NSO_2), 1173 (NSO_2); ^{1}H NMR (400 MHz, $CDCl_3$) δ 1.23 (t, $J = 7.4$ Hz, 3H), 2.35 (s, 3H), 2.87 (q, $J = 7.4$ Hz, 2H), 4.15 (s, 2H), 7.13 (dd, $J = 8.0$, 8.0 Hz, 1H), 7.24 (d, $J = 8.3$ Hz, 2H), 7.38 (d, $J = 8.0$ Hz, 1H), 7.60 (s, 1H), 7.75 (d, $J = 8.3$ Hz, 2H), 7.95 (d, $J = 8.0$ Hz, 1H); ^{13}C NMR (100 MHz, $CDCl_3$) δ 14.6, 21.6, 23.6, 40.9, 112.9, 114.5, 115.0, 125.7, 126.9 (2C), 127.8, 127.9, 128.6, 130.0 (2C), 134.8, 136.3, 145.3, 197.5. *Anal.* Calcd for $C_{19}H_{18}BrNO_3S_2$: C, 50.44; H, 4.01; N, 3.10. Found: C, 50.21; H, 4.01; N, 3.02.

5.1.7 3-Allyl-4-bromo-1H-indole (18)

The allylation of 4-bromoindole **17** was carried out according to the method of Tamaru [27]. To a stirred mixture of 4-bromoindole **17** (5.00 g, 25.5 mmol) in THF (65 mL) were added $Pd(PPh_3)_4$ (884 mg, 0.765 mmol), Et_3B (1.02 M solution in hexane; 7.5 mL, 7.65 mmol) and allyl alcohol (1.75 mL, 25.8 mmol) at room temperature under argon, and the mixture was stirred for 17 h at 50 °C. The mixture was concentrated under reduced pressure to give a brown oil, which was purified by flash chromatography over silica gel with *n*-hexane–EtOAc (8:1) to give **18** as a brown oil (5.24 g, 87% yield). All spectral data were in agreement with those reported by Tamaru [27].

5.1.8 3-Allyl-4-bromo-1-tosyl-1H-indole (19)

To a stirred solution of allylbromoindole **18** (5.24 g, 22.2 mmol), NaOH (2.66 g, 66.6 mmol) and *n*-Bu_4NHSO_4 (754 mg, 2.22 mmol) in CH_2Cl_2 (190 mL) was added TsCl (4.65 g, 24.4 mmol) at 0 °C. After stirring for 2.5 h at room temperature, 1,3-diaminopropane (1.11 mL, 13.3 mmol) and Et_3N (1.84 mL, 13.3 mmol) were added. The mixture was stirred for 2 h at this temperature and H_2O was added. The whole was extracted with EtOAc. The extract was washed with 1 N HCl, H_2O, and brine and dried over $MgSO_4$. Concentration under pressure gave a white solid, which was purified by column chromatography over silica gel with *n*-hexane–EtOAc (10:1) to give **19** (8.36 g, 96% yield). Recrystallization from *n*-hexane–EtOAc gave pure **19** as colorless crystals. All spectral data were in agreement with those reported by Hegedus [28].

5.1.9 2-(4-Bromo-1-tosyl-1H-indol-3-yl)acetaldehyde (8)

To a stirred mixture of **19** (700 mg, 1.79 mmol) and NMO (377 mg, 3.22 mmol) in THF/H_2O (14 mL, 3:1) was added OsO_4 (2.5 wt% *t*-BuOH, 0.912 mL, 0.090 mmol) at 0 °C. The mixture was stirred for 14 h at room temperature and quenched with saturated Na_2SO_3. After stirring for 15 min, the whole was extracted with EtOAc. The extract was washed with H_2O, brine and dried over $MgSO_4$. The filtrate was concentrated under reduced pressure to give a crude diol as a white amorphous solid, which was used without further purification. To a stirred solution of this diol in THF/H_2O (14 mL, 3:1) was added $NaIO_4$ (1.53 g, 7.16 mmol) at room temperature. After stirring for 2.5 h at this temperature, the mixture was diluted with EtOAc. The organic phase was separated and washed with H_2O, brine and dried over $MgSO_4$. The filtrate was concentrated under reduced pressure to give an oily residue, which was purified by flash chromatography over silica gel with *n*-hexane–EtOAc (4:1) to give **8** as a yellow oil (606 mg, 86% yield): IR (neat): 1725 (C=O), 1371 (NSO_2), 1172 (NSO_2); ^{1}H NMR (500 MHz, $CDCl_3$) δ 2.35 (s, 3H), 4.06 (s, 2H), 7.15 (dd, $J = 8.6$, 8.6 Hz, 1H), 7.24 (d, $J = 8.0$ Hz, 2H), 7.37 (d, $J = 8.6$ Hz, 1H), 7.58 (s, 1H), 7.76 (d, $J = 8.0$ Hz, 2H), 7.97 (d, $J = 8.6$ Hz, 1H), 9.87 (s, 1H); ^{13}C NMR (125 MHz, $CDCl_3$) δ 21.5, 40.8, 113.0, 113.8, 114.2, 125.8, 126.9 (3C), 127.8, 128.3, 130.0

(2C), 134.8, 136.2, 145.5, 198.6. HRMS (FAB) calcd $C_{17}H_{15}BrNO_3S$: $[M+H]^+$, 391.9956; found: 391.9954.

5.1.10 *(R)-1-(4-Bromo-1-tosyl-1H-indol-3-yl)-4-[(2R,5S)-2-phenyl-3-tosyl-1,3-oxazinan-5-yl]but-3-yn-2-ol (20a) and Its (S)-Isomer (20b)*

To a stirred mixture of $NiCl_2$ (3.25 mg, 0.025 mmol) and $CrCl_2$ (324 mg, 2.51 mmol) in THF (6.3 mL) was added a solution of aldehyde **8** (246 mg, 0.63 mmol) and alkyne **13** (645 mg, 1.38 mmol) in THF (6.3 mL) at 0 °C under argon (Table 5.2, Entry 5). The mixture was stirred for 4.5 h at this temperature. The mixture was diluted with Et_2O and quenched with H_2O. The whole was extracted with EtOAc. The organic phase was separated and washed with saturated $Na_2S_2O_3$, brine and dried over $MgSO_4$. The filtrate was concentrated under reduced pressure to give an oily residue, which was purified by flash chromatography over silica gel with *n*-hexane–EtOAc (3:1) to give **20** as a pale yellow amorphous solid (414 mg, 90% yield, dr = 1:1).

5.1.11 *1-(4-Bromo-1-tosyl-1H-indol-3-yl)-4-[(2R,5S)-2-phenyl-3-tosyl-1,3-oxazinan-5-yl]-but-3-yn-2-one (16)*

To a stirred solution of alcohol **20** (2.64 g, 3.60 mmol) in CH_2Cl_2 (118 mL) was added Dess-Martin periodinane (3.37 g, 7.92 mmol) at 0 °C. After stirring for 20 min at this temperature, the mixture was allowed to warm to room temperature. The mixture was stirred for further 40 min at this temperature and quenched with saturated $Na_2S_2O_3$ and saturated $NaHCO_3$. The whole was extracted with EtOAc. The extract was washed with saturated $NaHCO_3$, brine and dried over $MgSO_4$, and concentrated under pressure to give an oily residue, which was purified by column chromatography over silica gel with *n*-hexane–EtOAc (3:1) to give **16** as a yellow amorphous solid (2.49 g, 95% yield): $[\alpha]_D^{28}$ −61.0 (*c* 1.17, $CHCl_3$); IR (neat): 2215 (C≡C), 1677 (C=O), 1375 (NSO_2), 1350 (NSO_2), 1171 (NSO_2); ^1H NMR (500 MHz, $CDCl_3$) δ 2.31–2.38 (m, 1H), 2.35 (s, 3H), 2.48 (s, 3H), 3.00 (dd, J = 14.6, 11.7 Hz, 1H), 3.35 (dd, J = 11.5, 10.9 Hz, 1H), 3.47–3.53 (m, 1H), 3.83 (dd, J = 14.6, 4.0 Hz, 1H), 4.03 (s, 2H), 6.66 (s, 1H), 7.08 (dd, J = 8.6, 8.6 Hz, 1H), 7.22–7.26 (m, 3H), 7.36–7.42 (m, 5H), 7.43–7.49 (m, 3H), 7.72 (d, J = 8.6 Hz, 2H), 7.82 (d, J = 8.0 Hz, 2H), 7.91 (d, J = 8.6 Hz, 1H); ^{13}C NMR (125 MHz, $CDCl_3$) δ 21.6, 21.7, 25.8, 42.0, 42.8, 62.1, 82.6, 83.0, 89.4, 112.9, 114.3 (2C), 125.8, 126.9 (2C), 127.0 (2C), 127.2, 127.5 (2C), 127.7, 128.3, 128.7, 129.3 (2C), 130.1 (4C), 134.6, 134.7, 136.1, 137.1, 144.3, 145.6, 183.8. HRMS (FAB) calcd $C_{36}H_{30}BrN_2O_6S_2$: $[M-H]^-$, 729.0734; found: 729.0734.

5.1.12 (R)-1-(4-Bromo-1-tosyl-1H-indol-3-yl)-4-[(2R,5S)-2-phenyl-3-tosyl-1,3-oxazinan-5-yl]but-3-yn-2-ol (20a)

A solution of (*R*)-Alpine–Borane (0.5 M in THF, 5.1 mL, 2.57 mmol) was slowly added to ketone **16** (628 mg, 0.858 mmol) at 0 °C under argon. The resulting solution was stirred for 32 h at room temperature. After the mixture was concentrated under reduced pressure, the residue was diluted with Et_2O (24 mL). Ethanolamine (0.194 mL, 3.22 mmol) was slowly added, producing a yellow precipitate, which was removed by filtration through Celite. The filtrate was concentrated under reduced pressure to give an oily residue, which was purified by flash chromatography over silica gel with *n*-hexane–EtOAc (3:1) to give **20a** as a pale yellow amorphous solid (539 mg, 86% yield, dr = > 95:5): $[\alpha]_D^{28}$ −59.1 (*c* 1.25, $CHCl_3$); IR (neat): 3507 (OH), 1374 (NSO_2), 1348 (NSO_2), 1171 (NSO_2); ^{1}H NMR (500 MHz, $CDCl_3$) δ 1.73 (d, J = 5.2 Hz, 1H), 2.28–2.36 (m, 1H), 2.35 (s, 3H), 2.47 (s, 3H), 3.08 (dd, J = 14.3, 12.0 Hz, 1H), 3.20–3.24 (m, 2H), 3.47 (dd, J = 11.5, 11.5 Hz, 1H), 3.59–3.64 (m, 1H), 3.93 (dd, J = 14.3, 4.6 Hz, 1H), 4.53–4.60 (m, 1H), 6.68 (s, 1H), 7.09 (dd, J = 8.6, 8.6 Hz, 1H), 7.22 (d, J = 8.6 Hz, 2H), 7.30 (d, J = 8.6 Hz, 1H), 7.35–7.39 (m, 1H), 7.37 (d, J = 8.6 Hz, 2H), 7.43–7.46 (m, 5H), 7.70 (d, J = 8.6 Hz, 2H), 7.86 (d, J = 8.6 Hz, 2H), 7.93 (d, J = 8.6 Hz, 1H); ^{1}H NMR [500 MHz, $(CD_3)_2SO$] δ 2.08–2.17 (m, 1H), 2.31 (s, 3H), 2.43 (s, 3H), 2.86 (dd, J = 14.3, 12.6 Hz, 1H), 3.01–3.16 (m, 3H), 3.44 (dd, J = 11.7, 3.7 Hz, 1H), 3.79–3.86 (m, 1H), 4.37–4.42 (m, 1H), 5.48 (d, J = 5.7 Hz, 1H), 6.56 (s, 1H), 7.17 (dd, J = 8.6, 8.6 Hz, 1H), 7.27–7.33 (m, 3H), 7.34 (d, J = 8.6 Hz, 2H), 7.42 (dd, J = 8.0, 8.0 Hz, 1H), 7.46–7.53 (m, 4H), 7.66 (s, 1H), 7.78 (d, J = 8.0 Hz, 2H), 7.86 (d, J = 8.6 Hz, 2H), 7.89 (d, J = 8.6 Hz, 1H); ^{13}C NMR (125 MHz, $CDCl_3$) δ 21.6 (2C), 25.7, 34.3, 43.9, 62.2, 63.2, 81.5, 83.0, 84.0, 112.9, 114.3, 117.3, 125.5, 126.8 (3C), 127.1 (2C), 127.5 (2C), 127.9, 128.4, 128.5, 129.2 (2C), 130.0 (4C), 134.8, 135.1, 136.3, 137.5, 144.0, 145.4. HRMS (FAB) calcd $C_{36}H_{32}BrN_2O_6S_2$: $[M–H]^-$, 731.0891; found: 731.0889.

5.1.13 (2R,5S)-5-[(R)-4-(4-Bromo-1-tosyl-1H-indol-3-yl)buta-1,2-dienyl]-2-phenyl-3-tosyl-1,3-oxazinane (21a)

To a stirred solution of PPh_3 (766 mg, 2.92 mmol) in THF (10 mL) was added diethyl azodicarboxylate (40% solution in toluene, 1.33 mL, 2.92 mmol) at −15 °C under argon. After stirring for 5 min at this temperature, a solution of propargylic alcohol **20a** (535 mg, 0.729 mmol) in THF (8.0 mL) was added to the reaction mixture, followed 5 min later by the addition of a solution of *o*-nitrobenzenesulfonyl hydrazide (634 mg, 2.92 mmol) in THF (9.0 mL) at −15 °C. After stirring for 2.5 h at this temperature, the mixture was allowed to warm to room temperature and stirred for further 5 h at this temperature. Concentration under reduced pressure gave an oily residue, which was purified by flash chromatography over silica gel with *n*-hexane–EtOAc (4:1) to give **21a** as a pale yellow amorphous solid (404 mg, 77% yield, dr = 94:6): $[\alpha]_D^{28} -87.0$ (*c* 1.05, $CHCl_3$); IR (neat): 1964 (C=C=C), 1378 (NSO_2), 1343 (NSO_2), 1172 (NSO_2); ^1H NMR (500 MHz, $CDCl_3$) δ 1.94–2.02 (m, 1H), 2.34 (s, 3H), 2.46 (s, 3H), 2.86 (dd, J = 14.9, 12.0 Hz, 1H), 3.25 (dd, J = 11.5, 11.5 Hz, 1H), 3.39–3.44 (m, 1H), 3.49 (ddd, J = 16.2, 6.4, 2.1 Hz, 1H), 3.56 (ddd, J = 16.2, 6.3, 2.1 Hz, 1H), 3.75 (dd, J = 14.9, 4.6 Hz, 1H), 4.55–4.61 (m, 1H), 5.31–5.38 (m, 1H), 6.64 (s, 1H), 7.12 (dd, J = 8.0, 8.0 Hz, 1H), 7.22 (d, J = 8.6 Hz, 2H), 7.30 (s, 1H), 7.32–7.37 (m, 4H), 7.40–7.46 (m, 4H), 7.71 (d, J = 8.6 Hz, 2H), 7.85 (d, J = 8.6 Hz, 2H), 7.94 (d, J = 8.0 Hz, 1H); ^{13}C NMR (125 MHz, $CDCl_3$) δ 21.5, 21.6, 26.2, 31.8, 44.6, 64.2, 83.0, 88.9, 92.0, 112.9, 114.4, 121.4, 125.0, 125.5, 126.8 (2C), 127.1 (2C), 127.5 (2C), 127.7, 128.3, 128.5, 129.0 (2C), 129.9 (2C), 130.0 (2C), 134.9, 135.6, 136.4, 137.8, 143.7, 145.2, 204.4. HRMS (FAB) calcd $C_{36}H_{32}BrN_2O_5S_2$: $[M–H]^-$, 715.0941; found: 715.0941.

5.1.14 N-[(2S,4R)-6-(4-Bromo-1-tosyl-1H-indol-3-yl)-2-(hydroxymethyl)hexa-3,4-dienyl] -4-methylbenzenesulfon-amide (5a)

To a stirred mixture of **21a** (400 mg, 0.557 mmol, dr = 94:6) in $MeOH/CH_2Cl_2$ (20 mL, 1:1) was added *p*-toluenesulfonic acid monohydrate (159 mg, 0.836 mmol) at room temperature. After stirring for 3.5 h at 50 °C, concentration under reduced pressure gave an oily residue. The residue was dissolved in EtOAc and washed with saturated $NaHCO_3$, brine and dried over $MgSO_4$. The filtrate was concentrated under reduced pressure to give an oily residue, which was purified by flash chromatography over silica gel with *n*-hexane–EtOAc (3:2) to give **5a** as a white amorphous solid (299 mg, 85% yield, dr = 94:6): $[\alpha]_D^{28} -50.6$ (*c* 1.15, $CHCl_3$); IR (neat): 3313 (OH), 1963 (C=C=C), 1413 (NSO_2), 1372 (NSO_2), 1172 (NSO_2), 1157 (NSO_2); ^1H NMR (500 MHz, $CDCl_3$) δ 1.77 (t, J = 6.3 Hz, 1H), 2.21–2.29 (m, 1H), 2.35 (s, 3H), 2.43 (s, 3H), 2.79–2.90 (m, 2H), 3.39 (ddd, J = 10.3, 6.3, 5.7 Hz, 1H), 3.49 (ddd, J = 10.3, 6.3, 4.5 Hz, 1H), 3.56–3.69 (m, 2H), 4.68 (t, J = 6.6 Hz, 1H), 4.92–4.97 (m, 1H), 5.46–5.52 (m, 1H), 7.09 (dd, J = 8.0, 8.0 Hz, 1H), 7.23 (d, J = 8.0 Hz, 2H), 7.31 (d, J = 8.0 Hz, 2H), 7.35 (d, J = 8.0 Hz, 1H), 7.42 (s, 1H), 7.71 (d, J = 8.0 Hz, 2H), 7.73 (d, J = 8.0 Hz, 2H), 7.92 (d, J = 8.0 Hz, 1H); ^{13}C NMR (125 MHz, $CDCl_3$) δ 21.5 (2C), 26.2, 40.7, 44.2, 63.4, 90.3, 92.0, 112.8, 114.5, 121.3, 125.2, 125.5, 126.8 (2C), 127.0 (2C), 127.7, 128.6, 129.7 (2C), 130.0 (2C), 134.8, 136.4, 136.9, 143.4, 145.3, 204.7. HRMS (FAB) calcd $C_{29}H_{28}BrN_2O_5S_2$: $[M–H]^-$, 627.0628; found: 627.0627.

5.1.15 Determination of Relative Configuration of the Allenamide 5a: Synthesis of the Authentic Sample (±)-5a by Desilylation of the Known Allenamide (±)-22a

To a stirred solution of (±)-**22a** (2.6 mg, 0.0033 mmol) in THF (0.30 mL) was added TBAF (1.00 M solution in THF; 33 μL, 0.033 mmol) at 0 °C. The mixture was stirred for 30 min at room temperature. Concentration under reduced pressure gave an oily residue, which was purified by flash chromatography over silica gel with *n*-hexane–EtOAc (3:2) to give (±)-**5a** as a white amorphous solid (1.9 mg, 91% yield).

5.1.16 (S)-1-(4-Bromo-1-tosyl-1H-indol-3-yl) -4-[(2R,5S)-2-phenyl-3-tosyl-1,3-oxazinan-5-yl] but-3-yn-2-ol (20b)

A solution of (*S*)-Alpine–borane (0.5 M in THF, 0.82 mL, 0.137 mmol) was slowly added to ketone **16** (100 mg, 0.410 mmol) at 0 °C under argon. The resulting solution was stirred for 33 h at room temperature. After the mixture was concentrated under reduced pressure, the residue was diluted with Et_2O (3.8 mL). Aminoethanol (0.031 mL, 0.514 mmol) was slowly added, producing a yellow precipitate which was removed by filtration through Celite. The filtrate was concentrated under reduced pressure to give an oily residue, which was purified by flash chromatography over silica gel with *n*-hexane–EtOAc (3:1) to give **20b** as a pale yellow amorphous solid (69.5 mg, 69% yield, dr = > 95:5): $[\alpha]_D^{28}$ −38.2 (*c* 0.84, $CHCl_3$); IR (neat): 3501 (OH), 1373 (NSO_2), 1347 (NSO_2), 1170 (NSO_2); ^{1}H NMR (500 MHz, $CDCl_3$) δ 1.73 (d, *J* = 5.2 Hz, 1H), 2.28–2.36 (m, 1H), 2.35 (s, 3H), 2.46 (s, 3H), 3.10 (dd, *J* = 14.3, 12.0 Hz, 1H), 3.18 (dd, *J* = 14.6, 6.0 Hz, 1H), 3.25 (dd, *J* = 14.6, 6.6 Hz, 1H), 3.44 (dd, *J* = 11.5, 11.5 Hz, 1H), 3.58 (dd, *J* = 11.5, 4.0 Hz, 1H), 3.95 (dd, *J* = 14.3, 2.9 Hz, 1H), 4.54–4.60 (m, 1H), 6.68 (s, 1H), 7.09 (dd, *J* = 8.0, 8.0 Hz, 1H), 7.21 (d, *J* = 8.0 Hz, 2H), 7.30 (d, *J* = 8.0 Hz, 1H), 7.35–7.41 (m, 3H), 7.43–7.46 (m, 5H), 7.69 (d, *J* = 8.0 Hz, 2H), 7.87 (d, *J* = 8.0 Hz, 2H), 7.93 (d, *J* = 8.0 Hz, 1H); ^{1}H NMR [500 MHz, $(CD_3)_2SO$] δ 2.06–2.15 (m, 1H), 2.31 (s, 3H), 2.42 (s, 3H), 2.85 (dd, *J* = 14.9, 12.0 Hz, 1H), 3.01 (dd, *J* = 10.9, 7.4 Hz, 1H), 3.11–3.18 (m, 2H), 3.49 (dd, *J* = 10.9, 4.0 Hz, 1H), 3.81 (dd, *J* = 14.9, 4.6 Hz, 1H), 4.36–4.41 (m, 1H), 5.48 (d, *J* = 5.7 Hz, 1H), 6.57 (s, 1H), 7.18 (dd, *J* = 8.6, 8.6 Hz, 1H), 7.29–7.35 (m, 5H), 7.41 (dd, *J* = 8.0, 8.0 Hz, 1H), 7.45–7.53 (m, 4H), 7.68 (s, 1H), 7.78 (d, *J* = 8.6 Hz, 2H), 7.87 (d, *J* = 8.6 Hz, 2H), 7.89 (d, *J* = 8.6 Hz, 1H); ^{13}C NMR (125 MHz, $CDCl_3$) δ 21.6 (2C), 25.7, 34.3, 43.9, 62.2, 63.2, 81.4, 83.0, 84.0, 112.9, 114.3, 117.3, 125.4, 126.8 (3C), 127.1 (2C), 127.5 (2C), 127.9, 128.4,

128.5, 129.1 (2C), 130.0 (2C), 130.1 (2C), 134.8, 135.1, 136.2, 137.5, 144.0, 145.4. HRMS (FAB) calcd $C_{36}H_{32}BrN_2O_6S_2$: $[M-H]^-$, 731.0891; found: 731.0891.

5.1.17 *(2R,5S)-5-[(S)-4-(4-Bromo-1-tosyl-1H-indol-3-yl)buta-1,2-dienyl]-2-phenyl-3-tosyl-1,3-oxazinane (21b)*

By a procedure identical with that described for synthesis of **21a** from **20a**, the propargylic alcohol **20b** (135 mg, 0.184 mmol) was converted into **21b** as a pale yellow amorphous solid (80.5 mg, 61% yield, dr = 94:6): $[\alpha]_D^{28}$ +20.3 (*c* 0.78, $CHCl_3$); IR (neat): 1964 (C=C=C), 1374 (NSO_2), 1344 (NSO_2), 1172 (NSO_2); ^{1}H NMR (500 MHz, $CDCl_3$) δ 1.88–1.96 (m, 1H), 2.30 (s, 3H), 2.46 (s, 3H), 2.87 (dd, J = 14.6, 12.0 Hz, 1H), 3.10 (dd, J = 11.2, 11.2 Hz, 1H), 3.12–3.17 (m, 1H), 3.48 (ddd, J = 16.6, 6.3, 3.0 Hz, 1H), 3.60 (ddd, J = 16.6, 6.9, 2.9 Hz, 1H), 3.75 (dd, J = 14.6, 4.3 Hz, 1H), 4.41–4.46 (m, 1H), 5.32–5.38 (m, 1H), 6.63 (s, 1H), 7.12 (dd, J = 8.6, 8.6 Hz, 1H), 7.20 (d, J = 8.6 Hz, 2H), 7.31–7.38 (m, 5H), 7.39–7.45 (m, 4H), 7.70 (d, J = 8.6 Hz, 2H), 7.82 (d, J = 8.6 Hz, 2H), 7.95 (d, J = 8.6 Hz, 1H); ^{13}C NMR (125 MHz, $CDCl_3$) δ 21.5, 21.6, 26.1, 32.0, 44.4, 64.3, 83.0, 89.0, 92.2, 112.9, 114.4, 121.3, 125.1, 125.5, 126.8 (2C), 127.1 (2C), 127.5 (2C), 127.8, 128.3, 128.6, 129.0 (2C), 129.9 (4C), 134.9, 135.6, 136.5, 137.8, 143.7, 145.2, 204.6. HRMS (FAB) calcd $C_{36}H_{32}BrN_2O_5S_2$: $[M-H]^-$, 715.0941; found: 715.0938.

5.1.18 *N-[(2S,4S)-6-(4-Bromo-1-tosyl-1H-indol-3-yl)-2-(hydroxymethyl)hexa-3,4-dienyl]-4-methylbenzenesulfon-amide (5b)*

By a procedure identical with that described for synthesis of **5a** from **21a**, the allene **21b** (77.7 mg, 0.108 mmol, dr = 94:6) was converted into **5b** as a white amorphous solid (58.5 mg, 86% yield, dr = 94:6): $[\alpha]_D^{28}$ +48.8 (*c* 1.57, $CHCl_3$); IR (neat): 3292 (OH), 1965 (C=C=C), 1412 (NSO_2), 1372 (NSO_2), 1172 (NSO_2), 1157 (NSO_2); ^{1}H NMR (500 MHz, $CDCl_3$) δ 1.85–1.91 (m, 1H), 2.17–2.26 (m, 1H), 2.35 (s, 3H), 2.42 (s, 3H), 2.88 (ddd, J = 12.0, 6.3, 5.7 Hz, 1H), 2.96 (ddd, J = 12.0, 6.3, 5.7 Hz, 1H), 3.33–3.46 (m, 2H), 3.59 (ddd, J = 16.5, 5.7, 2.9 Hz, 1H), 3.66 (ddd, J = 16.5, 6.4, 2.1 Hz, 1H), 4.79 (t, J = 6.3 Hz, 1H), 4.87–4.92 (m, 1H), 5.45–5.52 (m, 1H), 7.10 (dd, J = 8.6, 8.6 Hz, 1H), 7.23 (d, J = 8.6 Hz, 2H), 7.30 (d, J = 8.6 Hz, 2H), 7.32 (d, J = 8.6 Hz, 1H), 7.42 (s, 1H), 7.72 (d, J = 8.6 Hz, 2H), 7.73 (d, J = 8.6 Hz, 2H), 7.94 (d, J = 8.6 Hz, 1H);

^{13}C NMR (125 MHz, $CDCl_3$) δ 21.5, 21.6, 26.1, 40.8, 44.1, 63.5, 90.3, 92.0, 112.9, 114.4, 121.3, 125.2, 125.5, 126.9 (2C), 127.0 (2C), 127.7, 128.6, 129.8 (2C), 130.0 (2C), 134.9, 136.5, 136.9, 143.5, 145.3, 204.8. HRMS (FAB) calcd $C_{29}H_{28}BrN_2O_5S_2$: $[M–H]^-$, 627.0628; found: 627.0630.

5.1.19 [(6aR,9S)-4,7-Ditosyl-4,6,6a,7,8,9 -hexahydroindolo[4,3-fg]quinolin-9-yl]methanol (4a)

To a stirred mixture of allenamide **5a** (248 mg, 0.39 mmol, dr = 94:6) in DMF (8.0 mL) were added $Pd(PPh_3)_4$ (22.8 mg, 0.020 mmol) and K_2CO_3 (162 mg, 1.17 mmol) at room temperature under argon, and the mixture was stirred for 2.5 h at 100 °C. Concentration under reduced pressure gave an oily residue. The residue was dissolved in EtOAc and washed with saturated NH_4Cl, H_2O, and brine and dried over $MgSO_4$. The filtrate was concentrated under reduced pressure to give a brown oil, which was purified by flash chromatography over silica gel with *n*-hexane–EtOAc (1:1), followed by flash chromatography over Chromatorex® with *n*-hexane–EtOAc (1:1–1:2) to give **4a** as a pale brown amorphous solid (162 mg, 76% yield, dr = 92:8). The pure diastereomer **4a** was isolated by PTLC with *n*-hexane–EtOAc–MTBE (1:1:1): $[\alpha]_D^{28} -129.1$ (*c* 0.38, $CHCl_3$); IR (neat): 3523 (OH), 1597 (C=C), 1357 (NSO_2), 1342 (NSO_2), 1178 (NSO_2), 1155 (NSO_2); ^{1}H NMR (500 MHz, $CDCl_3$) δ 2.26–2.33 (m, 1H), 2.35 (s, 3H), 2.40 (s, 3H), 2.87–2.96 (m, 2H), 3.31 (dd, J = 14.3, 5.2 Hz, 1H), 3.51–3.63 (m, 2H), 4.08 (dd, J = 14.3, 5.2 Hz, 1H), 4.67–4.72 (m, 1H), 6.13 (s, 1H), 7.17 (d, J = 8.0 Hz, 1H), 7.20–7.29 (m, 6H), 7.68 (d, J = 8.0 Hz, 2H), 7.77 (d, J = 8.0 Hz, 2H), 7.79 (d, J = 8.0 Hz, 1H); ^{13}C NMR (125 MHz, $CDCl_3$) δ 21.5 (2C), 29.7, 37.1, 42.3, 53.4, 64.2, 112.8, 115.6, 117.5, 120.5, 123.7, 125.7, 126.8 (4C), 128.3, 129.8 (2C), 129.9 (2C), 130.1, 133.4, 134.1, 135.4, 138.0, 143.4, 144.9; HRMS (FAB) calcd $C_{29}H_{29}N_2O_5S_2$: $[M+H]^+$, 549.1518; found: 549.1516.

5.1.20 Determination of Relative Configuration of the Alcohol 4a: Synthesis of the Authentic Sample (±)-4a by Desilylation of the Known Tetracyclic Indole (±)-23a

To a stirred solution of (±)-**23a** (2.6 mg, 0.0037 mmol) in THF (0.30 mL) was added TBAF (1.00 M solution in THF; 37 μL, 0.037 mmol) at 0 °C. The mixture was stirred for 30 min at room temperature. Concentration under reduced pressure gave an oily residue, which was purified by flash chromatography over silica gel with *n*-hexane–EtOAc (1:1) to give (±)-**4a** as a white amorphous solid (2.2 mg, quant.).

5.1.21 *[(6aS,9S)-4,7-Ditosyl-4,6,6a,7,8,9-hexahydroindolo[4,3-fg]quinolin-9-yl]methanol (4b)*

By a procedure identical with that described for synthesis of **4a** from **5a**, the allenamide **5b** (25 mg, 0.040 mmol, dr = 94:6) was converted into **4b** as a pale brown amorphous solid (9.4 mg, 43% yield, dr = 69:31). The pure diastereomer **4b** was isolated by PTLC with *n*-hexane–EtOAc–MTBE (1:1:1): $[\alpha]_D^{28}$ +6.0 (*c* 0.19, $CHCl_3$); IR (neat): 3560 (OH), 1597 (C=C), 1359 (NSO_2), 1335 (NSO_2), 1178 (NSO_2), 1155 (NSO_2); ^1H NMR (500 MHz, $CDCl_3$) δ 2.35 (s, 3H), 2.35–2.38 (m, 1H), 2.46 (s, 3H), 2.55–2.61 (m, 1H), 2.72 (ddd, J = 14.3, 12.0, 1.7 Hz, 1H), 3.00 (dd, J = 14.3, 5.2 Hz, 1H), 3.08 (dd, J = 13.7, 2.9 Hz, 1H), 3.52 (ddd, J = 12.0, 10.3, 5.7 Hz, 1H), 3.66 (ddd, J = 12.0, 7.4, 5.2 Hz, 1H), 4.09 (d, J = 13.7 Hz, 1H), 4.77–4.82 (m, 1H), 6.30 (d, J = 5.7 Hz, 1H), 7.14 (d, J = 1.7 Hz, 1H), 7.17–7.30 (m, 4H), 7.35 (d, J = 8.6 Hz, 2H), 7.74 (d, J = 8.6 Hz, 2H), 7.77 (d, J = 8.6 Hz, 1H), 7.78 (d, J = 8.6 Hz, 2H); ^{13}C NMR (125 MHz, $CDCl_3$) δ 21.5 (2C), 27.3, 38.7, 39.8, 53.2, 61.6, 112.8, 115.8, 117.0, 120.5, 122.5, 125.9, 126.7 (4C), 128.0, 129.9 (4C), 130.2, 133.3, 134.8, 135.4, 138.6, 143.6, 144.9; HRMS (FAB) calcd $C_{29}H_{29}N_2O_5S_2$: $[M+H]^+$, 549.1518; found: 549.1519.

5.1.22 *(+)-Isolysergol (3)*

To a stirred solution of **4a** (20 mg, 0.036 mmol, dr = 92:8) in THF (0.50 mL) was added sodium naphthalenide (0.67 M solution in THF; 0.82 mL, 0.55 mmol) at −78 °C under argon. The mixture was stirred for 10 min at this temperature and quenched with saturated NH_4Cl. The mixture was made basic with saturated $NaHCO_3$. The whole was extracted with EtOAc. The extract was washed with brine and dried over $MgSO_4$. Concentration of the filtrate under reduced pressure gave a crude amine, which was used without further purification. To a stirred solution of this amine in MeOH (2.6 mL) were added formalin (0.028 mL, 0.36 mmol), $NaBH_3CN$ (22.6 mg, 0.36 mmol) and AcOH (47 μL) at room temperature. After stirring for 1 h at this temperature, the mixture was quenched with saturated $NaHCO_3$. After MeOH was removed under reduced pressure, the whole was extracted with EtOAc. The extract was washed with saturated $NaHCO_3$, brine and dried over $MgSO_4$. The filtrate was concentrated under reduced pressure to give a yellow solid, which was purified by PTLC (Chromatorex®) with *n*-hexane–EtOAc (1:10) to give isolysergol (**3**) as a pale brown solid (4.2 mg, 46% yield, 99% ee [HPLC, Chiralcel-OD column eluting with 10:90 hexane/*i*PrOH at 0.5 mL/min, t_1 = 9.58 min (minor isomer), t_2 = 13.18 min (major isomer)] [4]). $[\alpha]_D^{28}$ +200.3 (*c* 0.37, pyridine) [lit. $[\alpha]_D^{20}$ +228 (*c* 0.40, pyridine)] [45]. All the spectral data were in agreement with those of the racemic sample as described Chap. 4.

5.1.23 Methyl (6aR,9S)-4,7-Ditosyl-4,6,6a,7,8,9-hexahydroindolo[4,3-fg]quinoline-9-carboxylate (24a)

To a stirred solution of **4a** (40 mg, 0.072 mmol, dr = 92:8) in CH_2Cl_2 (3.2 mL) was added Dess-Martin periodinane (124 mg, 0.29 mmol) at 0 °C. After stirring for 30 min at this temperature, the mixture was warming to room temperature. The mixture was stirred for further 1.5 h at this temperature and filtrated through a short pad of SiO_2 with EtOAc. The filtrate was concentrated under reduced pressure to give a crude aldehyde, which was used without further purification. To a stirred mixture of the crude aldehyde and 2-methylbut-2-ene (0.44 mL, 4.32 mmol) in a mixed solvent of THF (1.5 mL) and *t*-BuOH (1.5 mL) were added a solution of $NaClO_2$ (62.4 mg, 0.69 mmol) and NaH_2PO_4 (82.8 mg, 0.69 mmol) in H_2O (0.71 mL) at room temperature. After stirring for 1.5 h at room temperature, brine was added to the mixture. The whole was extracted with EtOAc. The extract was washed with brine and dried over $MgSO_4$. The filtrate was concentrated under reduced pressure to give a crude carboxylic acid, which was purified by flash chromatography over silica gel with a gradient solvent [*n*-hexane–EtOAc (1:2) to EtOAc–MeOH (9:1)]. To a stirred solution of this carboxylic acid in a mixed solvent of toluene (1.5 mL) and MeOH (0.79 mL) was added $TMSCHN_2$ (2.00 M solution in Et_2O; 0.36 mL, 0.72 mmol) at 0 °C. The mixture was stirred for 20 min at room temperature. Concentration under pressure gave an oily residue, which was purified by flash chromatography over silica gel with *n*-hexane–EtOAc (3:1) to give **24a** as a pale yellow amorphous solid (26.5 mg, 64% yield, dr = > 95:5). $[\alpha]_D^{25}$ −93.2 (*c* 0.95, $CHCl_3$). All the spectral data were in agreement with those of the racemic sample as described Chap. 4.

5.1.24 (+)-Methyl Isolysergate (25a) and (+)-Methyl Lysergate (25b)

The preparation of (+)-methyl isolysergate (**25a**) and (+)-methyl lysergate (**25b**) from **24a** was carried out according to the racemic synthesis as described Chap. 4: to a stirred solution of **24a** (26 mg, 0.045 mmol) in THF (1.4 mL) was added sodium naphthalenide (0.67 M solution in THF; 0.67 mL, 0.45 mmol) at −78 °C under argon. The mixture was stirred for 6 min at this temperature and quenched with saturated NH_4Cl. The mixture was made basic with saturated $NaHCO_3$. The whole was extracted with EtOAc. The extract was washed with brine and dried over $MgSO_4$. Concentration of the filtrate under reduced pressure gave a crude amine, which was used without further purification. To a stirred solution of this amine in MeOH (3.0 mL) were added AcOH (47 μL), $NaBH_3CN$ (14.1 mg, 0.23 mmol) and formalin (17.7 μL, 0.23 mmol) at room temperature. After stirring for 2 h at this temperature, the mixture was quenched with saturated $NaHCO_3$. The mixture was concentrated under pressure, and the whole was extracted with EtOAc. The extract was washed with saturated $NaHCO_3$ and brine, and dried over $MgSO_4$. The filtrate was concentrated under reduced pressure to give a yellow solid, which was purified by flash chromatography over silica gel with *n*-hexane–EtOAc (1:3–1:10) to give **25** as a yellow solid (8.2 mg, 65% yield, **a**:**b** = 33:67). All the spectral data were in agreement with those of the racemic sample as described Chap. 4.

5.1.25 (+)-Lysergol (2) and (+)-Isolysergol (3)

To a stirred solution of **25** (8.2 mg, 0.029 mmol, **a**:**b** = 33:67) in THF (0.5 mL) was added $LiAlH_4$ (5.5 mg, 0.145 mmol) at 0 °C. The mixture was stirred for 10 min at this temperature and quenched with saturated Na_2SO_4. The whole was extracted with EtOAc. The extract was washed with brine and dried over $MgSO_4$. The filtrate was concentrated under reduced pressure to give a yellow solid, which was purified by PTLC (Chromatorex®) with EtOAc–MeOH (9:1) to give isolysergol (**3**) as a pale brown solid (1.8 mg, 24% yield, 97% ee), and lysergol (**2**) as a pale brown solid (3.6 mg, 49% yield, 98% ee [HPLC, Chiralcel-OD column eluting with 70:30 hexane/EtOH at 0.5 mL/min, t_1 = 13.19 min (minor isomer), t_2 = 17.64 min (major isomer)]). $[\alpha]_D^{26}$+40.9 (*c* 0.32, pyridine) [lit. $[\alpha]_D^{20}$+54 (*c* 0.40, pyridine)] [45]. All the spectral data were in agreement with those of the racemic sample as described Chap. 4.

5.1.26 (+)-Lysergic Acid (1)

The preparation of lysergic acid (**1**) was carried out according to the method of Szántay [1]: to solution of diastereomixture of **25** (20.0 mg, 0.071 mmol, **a**:**b** = 33:67) in EtOH (0.69 mL) was added 1 N NaOH (0.69 mL). The reaction mixture was stirred for 2 h at 35 °C. 0.1 N HCl solution was used to carefully adjust the pH to 6.2 and stirred at 0 °C for further 2 h while a solid material was formed. The precipitate was filtered off and washed with cold water and acetone to give (+)-lysergic acid (**1**) as a pale brown solid (10.2 mg, 54% yield): mp 220–223 °C dec. (lit. mp. 230–240 °C dec.) [1]; $[\alpha]_D^{26}$+36.1 (*c* 0.14, pyridine) [lit. $[\alpha]_D^{20}$+40 (*c* 0.50, pyridine)] [1]. All the spectral data were in agreement with those of the racemic sample as described Chap. 4.

5.1.27 Determination of Optical Purity of Lysergic Acid (1)

To a stirred suspension of lysergic acid (**1**) (2.5 mg, 0.0093 mmol) in a mixed solvent of EtOH (0.5 mL) and benzene (0.25 mL) was added $TMSCHN_2$ (2.00 M solution in Et_2O; 0.047 mL, 0.093 mmol) at 0 °C. The mixture was stirred for 10 min at room temperature. Concentration under pressure gave an oily residue, which was purified by flash chromatography over silica gel with *n*-hexane–EtOAc (1:3–1:10) to give **25** as a pale yellow solid (2.5 mg, 95% yield, **a**:**b** = 15:85, > 95% ee (**25a**), 96% ee (**25b**) [HPLC, Chiralcel-OD column eluting with 80:20 hexane/EtOH at 0.5 mL/min, t_1 = 16.88 min (methyl lysergate, minor isomer), t_2 = 18.67 min (methyl isolysergate, minor isomer), t_3 = 25.08 min (methyl lysergate, major isomer), t_4 = 27.54 min (methyl isolysergate, major isomer)]).

References

1. Moldvai I, Temesvári-Major E, Incze M, Szentirmay É, Gács-Baitz E, Szántay C (2004) J Org Chem 69:5993–6000
2. Inoue T, Yokoshima S, Fukuyama T (2009) Heterocycles 79:373–378
3. Kurosawa T, Isomura M, Tokuyama H, Fukuyama T (2009) Synlett 775–777
4. Deck JA, Martin SF (2010) Org Lett 12:2610–2613
5. Weibel J-M, Blanc A, Pale P (2008) Chem Rev 108:3149–3173
6. Álvarez-Corral M, Muñoz-Dorado M, Rodríguez-García I (2008) Chem Rev 108:3174–3198
7. Patil NT, Yamamoto Y (2008) Chem Rev 108:3395–3442
8. Widenhoefer RA, Han X (2006) Eur J Org Chem 4555–4563
9. Shen HC (2008) Tetrahedron 64:3885–3903
10. Muzart J (2008) Tetrahedron 64:5815–5849
11. Widenhoefer RA (2008) Chem Eur J 14:5382–5391
12. Krause N, Belting V, Deutsch C, Erdsack J, Fan H-T, Gockel B, Hoffmann-Röder A, Morita N, Volz F (2008) Pure Appl Chem 80:1063–1069
13. Li Z, Brouwer C, He C (2008) Chem Rev 108:3239–3265
14. Bongers N, Krause N (2008) Angew Chem Int Ed 47:2178–2181
15. Arredondo VM, McDonald FE, Marks TJ (1998) J Am Chem Soc 120:4871–4872
16. Arredondo VM, Tian S, McDonald FE, Marks TJ (1999) J Am Chem Soc 121:3633–3639
17. Ohno H, Kadoh Y, Fujii N, Tanaka T (2006) Org Lett. 8:947–950
18. Myers AG, Zheng B (1996) J Am Chem Soc 118:4492–4493
19. Ohno H, Hamaguchi H, Tanaka T (2000) Org Lett 2:2161–2163
20. Ohno H, Hamaguchi H, Tanaka T (2001) J Org Chem 66:1867–1875
21. Garner P (1984) Tetrahedron Lett 25:5855–5858
22. Campbell AD, Raynham TM, Taylor RJK (1998) Synthesis 1707–1709
23. Ohno H, Hamaguchi H, Ohata M, Kosaka S, Tanaka T (2004) J Am Chem Soc 126: 8744–8754
24. Hofmeister H, Annen K, Laurent H, Wiechert R (1984) Angew Chem Int Ed Engl 23: 727–729
25. Tokuyama H, Miyazaki T, Yokoshima S, Fukuyama T (2003) Synlett 1512–1514
26. Somei M, Kizu K, Kunimoto M, Yamada F (1985) Chem Pharm Bull 33:3696–3708

27. Kimura M, Futamata M, Mukai R, Tamaru Y (2005) J Am Chem Soc 127:4592–4593
28. Harrington PJ, Hegedus LS (1984) J Org Chem 49:2657–2662
29. Imamoto T, Kusumoto T, Yokoyama M (1982) J Chem Soc Chem Commun 1042–1044
30. Takita R, Yakura K, Ohshima T, Shibasaki M (2005) J Am Chem Soc 127:13760–13761
31. Lu G, Li X, Chan WL, Chan ASC (2002) Chem Commun 172
32. Gao G, Moore D, Xie R-G, Pu L (2002) Org Lett 4:4143–4146
33. Evans PA, Cui J, Gharpure SJ, Polosukhin A, Zhang H-R (2003) J Am Chem Soc 125: 14702–14703
34. Okude Y, Hirano S, Hiyama T, Nozaki H (1977) J Am Chem Soc 99:3179–3181
35. Jin H, Uenishi J, Christ WJ, Kishi Y (1986) J Am Chem Soc 108:5644–5646
36. Takai K, Tagashira M, Kuroda T, Oshima K, Utimoto K, Nozaki H (1986) J Am Chem Soc 108:6048–6050
37. Wan Z-K, Choi H-W, Kang F-A, Nakajima K, Demeke DK, Kishi Y (2002) Org Lett 4: 4431–4434
38. Choi H-W, Nakajima K, Demeke D, Kang F-A, Jun H-S, Wan Z-KK, Kishi Y (2002) Org Lett 4:4435–4438
39. Namba K, Kishi Y (2004) Org Lett 6:5031–5033
40. Midland MM, Tramontano A, Kazubski A, Graham RS, Tsai DJS, Cardinv DB (1984) Tetrahedron 40:1371–1380
41. Matsumura K, Hashiguchi S, Ikariya T, Noyori R (1997) J Am Chem Soc 119:8738–8739
42. Corey EJ, Bakshi RK, Shibata S (1987) J Am Chem Soc 109:5551–5553
43. Corey EJ, Bakshi RK (1990) Tetrahedron Lett 31:611–614
44. Corey EJ, Helal CJ (1998) Angew Chem Int Ed 37:1986–2012
45. Stoll A, Hofmann A, Schlientz W (1949) Helv Chim Acta 32:1947–1956
46. Smith S, Timmis GM (1936) J Chem Soc 1440–1444
47. Moldvai I, Gács-Baitz E, Temesvári-Major E, Russo L, Pápai I, Rissanen K, Szárics É, Kardos J, Szántay C (2007) Heterocycles 71:1075–1095

Chapter 6
Conclusions

1. Total synthesis of pachastrissamine (jaspine B) through a novel palladium(0)-catalyzed bis-cyclization of bromoallenes bearing hydroxyl and benzamide groups as internal nucleophiles was achieved. The key feature of this synthetic route is the late-stage introduction of the long alkyl side chain into the tetrahydrofuran ring at the C-2 position. Pachastrissamine derivatives with various alkyl groups were produced using different alkylation reagents.
2. A short synthetic route was developed for pachastrissamine with a 26% overall yield in seven steps from Garner's aldehyde. This synthesis was via palladium-catalyzed bis-cyclization of propargyl chlorides and carbonates. The cyclization reactivity was found to be dependent on the relative configuration of the benzamide and leaving groups, and on the nature of the leaving groups.
3. A novel palladium(0)-catalyzed domino cyclization of allenes with aryl halide and amino groups at both ends of internal allenes was also developed. This domino cyclization led to the formation of a sequential carbon–carbon bond and a carbon–nitrogen bond for construction of the core structure of ergot alkaloids. With this domino cyclization as the key step, total synthesis of (±)-lysergic acid, (±)-lysergol and (±)-isolysergol was achieved.
4. Enantioselective total synthesis of (+)-lysergol, (+)-isolysergol and (+)-lysergic acid was achieved by palladium(0)-catalyzed domino cyclization of the chiral amino allene. Enantiomerically pure amino allene for use as the cyclization precursor was synthesized via palladium/indium-mediated reductive coupling reaction of L-serine-derived ethynylaziridine with formaldehyde and Nozaki–Hiyama–Kishi (NHK) reaction. The palladium(0)-catalyzed cyclization allowed creation of the stereochemistry at C5 by transfer of the axial chirality of the allene to central chirality. This synthetic route afforded (+)-lysergic acid in a 4% overall yield in 15 steps from the ethynylaziridine.

S. Inuki, *Total Synthesis of Bioactive Natural Products by Palladium-Catalyzed Domino Cyclization of Allenes and Related Compounds*, Springer Theses,
DOI: 10.1007/978-4-431-54043-4_6, © Springer 2012

In summary, novel methodology has been developed for highly efficient construction of functionalized heterocycles by palladium-catalyzed domino/cascade cyclization of allenes and related compounds. This methodology was expanded to the total synthesis of bioactive natural products, pachastrissamine, lysergic acid, lysergol and isolysergol. These findings contribute to the development of a facile and efficient synthetic strategy for complex natural products and biologically active compounds.